Maths Out Loud
Year 1

by
Fran Mosley

BEAM
York St John

Acknowledgements

Jane Prothero and Woodlands Primary School, Leeds

Karen Holman and Paddox Primary School, Rugby

Heather Nixon and Gayhurst Primary School, Buckinghamshire

John Ellard and Kingsley Primary School, Northampton

Jackie Smith, Catherine Torr and Roberttown CE J & I School, Kirklees

Wendy Price and St Martin's CE Primary School, Wolverhampton

Helen Elis Jones, University of Wales, Bangor

Ruth Trundley, Devon Curriculum Services, Exeter

Trudy Lines and Bibury CE Primary School, Gloucestershire

Elaine Folen and St Paul's Infant School, Surrey

Jane Airey and Frith Manor Primary School, Barnet

Beverley Godfrey, South Wales Home Educators' Network

Kay Brunsdon and Gwyrosydd Infant School, Swansea

Keith Cadman, Wolverhampton Advisory Services

Helen Andrews and Blue Coat School, Birmingham

Oakridge Parochial School, Gloucestershire

The Islington BEAM Development Group

Published by BEAM Education
Maze Workshops
72a Southgate Road
London N1 3JT
Telephone 020 7684 3323
Fax 020 7684 3334
Email info@beam.co.uk
www.beam.co.uk
© Beam Education 2006
ISBN 1 903142 83 0
British Library Cataloguing-in-Publication Data
Data available
Edited by Ros Elphinstone and Marion Dill
Designed by Malena Wilson-Max
Photographs by Len Cross
Thanks to Rotherfield Primary School
Printed in England by Cromwell Press Ltd

Contents

Introduction

Language plays an important part in the learning of mathematics – especially oral language. Children's relationship to the subject, their grasp of it and sense of ownership all depend on discussion and interaction – as do the social relationships that provide the context for learning. A classroom where children talk about mathematics is one that will help build their confidence and transform their whole attitude to learning.

Why is speaking and listening important in maths?

- Talking is creative. In expressing thoughts and discussing ideas, children actually shape these ideas, make connections and hone their definitions of what words mean.
- You cannot teach what a word means – you can only introduce it, explain it, then let children try it out, misuse it, see when it works and how it fits with what they already know and, eventually, make it their own.
- Speaking and listening to other children involves and motivates children – they are more likely to learn and remember than when engaging silently with a textbook or worksheet.
- As you listen to children, you identify children's misconceptions and realise which connections (between bits of maths) they have not yet made.

How does this book help me include 'speaking and listening' in maths?

- The lessons are structured to use and develop oral language skills in mathematics. Each lesson uses one or more classroom techniques that foster the use of spoken language and listening skills.
- The grid on p15 shows those speaking and listening objectives, which are suitable for developing through the medium of mathematics. Each lesson addresses one of these objectives.
- The lessons draw on a bank of classroom techniques which are described on p8. These techniques are designed to promote children's use of speaking and listening in a variety of ways.

How does 'using and applying mathematics' fit in with these lessons?

- Many of the mathematical activities in this book involve problem solving, communication and reasoning, all key areas of 'using and applying mathematics' (U&A). Where this aspect of a lesson is particularly significant, this is acknowledged and expanded on in one of the 'asides' to the main lesson.

What about children with particular needs?

- For children who have impaired hearing, communication is particularly important, as it is all too easy for them to become isolated from their peers. Speaking and listening activities, even if adapted, simplified or supported by an assistant, help such children be a part of their learning community and to participate in the curriculum on offer.

- Children who speak English as an additional language benefit from speaking and listening activities, especially where these are accompanied by diagrams, drawings or the manipulation of numbers or shapes, which help give meaning to the language. Check that they understand the key words needed for the topic being discussed and, where possible, model the activity, paying particular attention to the use of these key words. Remember to build in time for thinking and reflecting on oral work.

- Differences in children's backgrounds affect the way they speak to their peers and adults. The lessons in this book can help children acquire a rich repertoire of ways to interact and work with others. Children who are less confident with written forms can develop confidence through speaking and listening.

- Gender can be an issue in acquiring and using speaking and listening skills. Girls may be collaborative and tentative, while boys sometimes can be more assertive about expressing their ideas. Address such differences by planning different groups, partners, classroom seating and activities. These lessons build on children's strengths and challenge them in areas where they are less strong.

What are the 'personal skills' learning objectives?

- There is a range of personal and social skills that children need to develop across the curriculum and throughout their school career. These include enquiry skills, creative thinking skills and ways of working with others. Some are particularly relevant to the maths classroom, and these are listed on the grid on p16.

What about assessment?

- Each lesson concludes with a section called 'Assessment for learning', which offers suggestions for what to look out for during the lesson and questions to ask in order to assess children's learning of all three learning objectives. There is also help on what may lie behind children's failure to meet these objectives and suggestions for teaching that might rectify the situation.

- Each section of four lessons includes a sheet of self-assessment statements to be printed from the accompanying CD-ROM and to be filled in at the end of each lesson or when all four are completed. Display the sheet and also give children their own copies. Then go through the statements, discussing and interpreting them as necessary. Ask children to complete their self-assessments with a partner they frequently work with. They should each fill in their own sheet, then look at it with their partner who adds their own viewpoint.

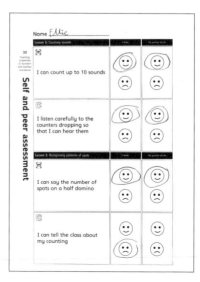

How can I make the best use of these lessons?

- Aim to develop a supportive classroom climate, where all ideas are accepted and considered, even if they may seem strange or incorrect. You will need to model this yourself in order for children to see what acceptance and open-mindedness look like.
- Create an ethos of challenge, where children are required to think about puzzles and questions.
- Slow down. Don't expect answers straight away when you ask questions. Build in thinking time where you do not communicate with the children, so that they have to reflect on their answers before making them. Expect quality rather than quantity.
- Model the language of discussion. Children who may be used to maths being either 'correct' or 'incorrect' need to learn by example what debate means. Choose a debating partner from the class, or work with another adult, and demonstrate uncertainty, challenge, exploration, questioning ...
- Tell children what they will be learning in the lesson. Each lesson concludes with an 'Assessment for learning' section offering suggestions for what to look out for to assess children's learning of all three learning objectives. Share these with the children at the start of the lesson to involve them in their own learning process.

How should I get the best out of different groupings?

- Get children used to working in a range of different groupings: pairs, threes or fours or as a whole class.
- Organise pairs in different ways on different occasions: familiar maths partners (who can communicate well); pairs of friends (who enjoy working together); children of differing abilities (who can learn something from each other); someone they don't know (to get them used to talking and listening respectfully to any other person).
- Give children working in pairs and groups some time for independent thought and work.
- Support pairs when they prepare to report back to the class. Go over with them what they have done or discovered and what they might say about this. Help them make brief notes – just single words or phrases – to remind them what they are going to say. If you are busy, ask an assistant or another child to take over your role. Then, when it comes to feedback time, support them by gentle probes or questions: "What did you do next?" or "What do your notes say?"

Classroom techniques used in this book

Ways of working

Peer tutoring
pairs of children

good for

This technique can benefit both the child who is being 'taught' and also the 'tutor' who develops a clearer understanding of what they themselves have learned and, in explaining it, can make new connections and solidify old ones. Children often make the best teachers, because they are close to the state of not knowing and can remember what helped them bridge the gap towards understanding.

how to organise it

'Peer tutoring' can work informally – children work in mixed ability pairs, and if one child understands an aspect of the work that the other doesn't, they work together in a tutor/pupil relationship to make sure the understanding is shared by both. Alternatively, you can structure it more formally. Observe children at work and identify those who are confident and accurate with the current piece of mathematics. Give them the title of 'Expert' and ask them to work with individuals needing support. Don't overuse this: the tutor has a right to work and learn at their own level, and tutoring others should only play a small part in their school lives.

Talking partners
pairs of children

good for

This technique helps children develop and practise the skills of collaboration in an unstructured way. Children can articulate their thinking, listen to one another and support each other's learning in a 'safe' situation.

how to organise it

Pairs who have previously worked together (for example, 'One between two', below) work together informally. The children in these pairs have had time to build up trust between them, and should have the confidence to tackle a new, less structured task. If you regularly use 'Talking partners', pairs of children will get used to working together. This helps them develop confidence, but runs the risk that children mutually reinforce their misunderstandings. In this case, changing partners occasionally can bring fresh life to the class by creating new meetings of minds.

One between two
pairs of children

good for

This technique helps children develop their skills of explaining, questioning and listening – behaviours that are linked to positive learning outcomes. Use it when the children have two or more problems or calculations to solve.

how to organise it

Pairs share a pencil (or calculator or other tool), and each assumes one of two roles: 'Solver' or 'Recorder'. (Supplying just one pencil encourages children to stay in role by preventing the Solver from making their own notes.)

The Solver has a problem and works through it out loud. The Recorder keeps a written record of what the Solver is doing. If the Solver needs something written down or a calculation done on the calculator, they must ask the Recorder to do this for them. If the Recorder is not sure of what the Solver is doing, they ask for further explanations, but do not engage in actually solving the problem. After each problem, children swap roles.

Introduce this way of working by modelling it yourself with a confident child partner: you talk through your own method of solving a problem, and the child records this thought process on the board.

Barrier games

pairs of children

good for

These techniques help children focus on spoken language rather than gesture or facial expression. The children must listen carefully to what is said, because they cannot see the person speaking.

how to organise it

Barrier games focus on giving and receiving instructions. Pairs of children work with a book or screen between them, so that they cannot see each other's work. The speaker gives information or instructions to the listener. The listener, in turn, asks questions to clarify understanding and gain information.

Eyes closed, eyes open

any number of children

good for

Depending on how this technique is used, it can either encourage children to listen carefully, because they cannot rely on visual checks, or to look carefully, because something was changed while they were not looking and they now need to identify this change.

how to organise it

Do this with the class: ask them to close their eyes while you count (slipping in a deliberate error) or drop coins into a tin. They must listen carefully to identify what you have done. Children then can do a similar activity in pairs.

Or tell children to close their eyes while you make one change in a sequence of numbers, pattern of shapes or some other structured set. When children open their eyes, they must spot what you have done and describe it or instruct you how to undo the change. Again, pairs can then carry on doing this independently.

Rotating roles

groups of various sizes

good for

Working in a small group to solve a problem encourages children to articulate their thinking and support each other's learning.

how to organise it

Careful structuring discourages individuals from taking the lead too often. Assign different roles to the children in the group: Chairperson, Reader, Recorder, Questioner, and so on. Over time, everyone has a turn at each role. You may wish to give children 'role labels' to remind them of their current role.

When you introduce this technique, model the role of chairperson in a group, with the rest of the class watching. Show how to include everyone and then discuss with the children what you have done, so as to make explicit techniques that they can use.

Discussion

Talking stick

any number of children

good for

Giving all children a turn at speaking and being listened to.

how to organise it

Provide the class with decorated sticks, which confer status on whoever holds them. Then, in a small or large group (or even the whole class), make it the rule that only the person holding the stick may speak, while the other children listen. You can use the stick in various ways: pass it around the circle; tell the child with the stick to pass it to whoever they want to speak next; have a chairperson who decides who will hold the stick next; ask the person with the stick to repeat what the previous person said before adding their own comments or ideas.

Tell your partner

pairs

good for

Whole-class question-and-answer sessions favour the quick and the confident and do not provide time and space for slower thinkers. This technique involves all children in answering questions and in discussion.

how to organise it

Do this in one of two ways:
- When you have just asked a question or presented an idea to think about, ask each child to turn to their neighbour or partner and tell them the answer. They then take turns to speak and to listen.

- Work less formally, simply asking children to talk over their ideas with a partner. Children may find this sharing difficult at first. They may not value talking to another child, preferring to talk to the teacher or not expressing their ideas at all. In this case, do some work on listening skills such as timing 'a minute each way' or repeating back to their partner what they have just said.

Devil's advocate
any number of children

good for

Statements – false or ambiguous as well as true – are often better than questions at provoking discussion.

how to organise it

In discussion with children, take the role of 'Devil's advocate', in which you make statements for them to agree or disagree with and to argue about.

To avoid confusing children by making false statements yourself, use a puppet that 'says' things that may not be true. Alternatively, explain that when you make statements with your hands behind your back, your fingers may be crossed and you may be saying things that are not true.

Reporting back

Heads or tails
pairs of children

good for

When pairs of children work together, one child may rely heavily on the other to make decisions and to communicate or one child may take over, despite the efforts of the other child to have a say. This technique encourages pairs to work together to understand something and helps prevent an uneven workload.

how to organise it

Invite pairs to the front of the class to explain their ideas or solutions. When they get to the front, ask them to nominate who is heads and who is tails, then toss a coin to decide which of them does the talking. They have one opportunity to 'ask a friend' (probably their partner). As all children in the class know that they may be chosen to speak in this way, because the toss of the coin could make either of them into the 'explainer', they are motivated to work with their partner to reach a common understanding. Assigning the choice of explainer to the toss of a coin stops children feeling that anyone is picking on them personally (do warn them in advance, though!).

Variation: If a pair of children has different ideas on a topic, ask both to offer explanations of each other's ideas.

Tell the class

individuals

good for

Encouraging children who lack skills or confidence to speak in front of the class.

how to organise it

In a plenary (or mini-plenary held during the course of a lesson), invite children to the front and support them in talking to the class by asking questions for them to answer. Gradually, experiences such as this can give children confidence to make their own statements.

Additional techniques

Below are some further classroom techniques that are referred to in the lessons in this book.

Ideas board

whole class

good for

An ideas board is a place where children display their work to the rest of the class informally and quickly. It also provides a useful place for you to record ideas and problems that you want children to think about.

how to organise it

The visual aspect of display is not a priority with an ideas board – it is more like a notice board where ideas and information can be shared. Make sure you remove items regularly to keep it fresh and up to date.

Chewing the fat

any number of children

good for

Leaving ideas or questions unresolved provides thoughtful children with the opportunity to extend their thinking and helps develop good habits. Many real mathematicians like to have problems to think about in odd moments, just as some people like crossword clues or chess moves to occupy their mind.

how to organise it

Sometimes end a lesson with ideas, problems or challenges for children to ponder in their own time as you may have run out of time or one of the children has come up with a question or an idea which can only be discussed the next day.

Charts

Classroom techniques

This chart shows which of the classroom techniques previously described are used in which lessons.

	COUNTING, PROPERTIES OF NUMBERS AND NUMBER SEQUENCES	PLACE VALUE AND ORDERING	CALCULATIONS	HANDLING DATA	MEASURES	SHAPE AND SPACE
	Lesson	Lesson	Lesson	Lesson	Lesson	Lesson
One between two		6	10	13		24
Talking partners	4		11	15		
Rotating roles					17	
Peer tutoring					20	
Eyes closed, eyes open	1	5				
Barrier games	3	7				21
Talking stick				14		23
Tell your partner		8			18	22
Devil's advocate				16		
Heads or tails			12		19	
Tell the class	2	9				

Speaking and listening skills

*This chart shows which speaking and listening skills are practised
in which lessons.*

	COUNTING, PROPERTIES OF NUMBERS AND NUMBER SEQUENCES	PLACE VALUE AND ORDERING	CALCULATIONS	HANDLING DATA	MEASURES	SHAPE AND SPACE
	Lesson	Lesson	Lesson	Lesson	Lesson	Lesson
Talk about shared work with a partner		8			18, 20	22
Reach a common understanding with a partner	4	7	11	16		21
Use precise language to explain ideas or give information						
Speak confidently in front of the class			9		19	
Give accurate instructions		5				24
Contribute to small-group and whole-class discussion	2		12	14, 15	17	
Listen and follow instructions accurately		6	10			
Listen to others and ask relevant questions						23
Listen with sustained concentration	1,3			13		

Personal skills

This chart shows which personal skills are practised in which lessons.

	COUNTING, PROPERTIES OF NUMBERS AND NUMBER SEQUENCES	PLACE VALUE AND ORDERING	CALCULATIONS	HANDLING DATA	MEASURES	SHAPE AND SPACE
	Lesson	Lesson	Lesson	Lesson	Lesson	Lesson
Organise work						
Identify stages in the process of fulfilling a task			11			
Check work		6	12			
Organise findings			10			
Work with others						
Discuss and agree ways of working				15		
Work cooperatively with others	3	7	9		18	21, 23
Overcome difficulties and recover from mistakes						24
Show awareness and understanding of others' needs	2	8			17	
Improve learning and performance						
Reflect on learning	1			14	20	
Critically evaluate own work		5			19	
Take pride in work				13		
Develop confidence in own judgements	4			16		22

Lessons

Counting, properties of numbers and number sequences

Learning objectives

	Lessons			
	1	2	3	4
Ⓜ Maths objectives				
count reliably at least ten objects	●	●		
recognise odd and even numbers			●	
describe and extend number sequences				●
Ⓢ Speaking and listening skills				
listen with sustained concentration	●		●	
contribute to whole-class discussion		●		
reach a common understanding with a partner				●
☺ Personal skills				
improve learning and performance: reflect on learning	●			
work with others: show awareness and understanding of others' needs		●		
work with others: work cooperatively with others			●	
improve learning and performance: develop confidence in own judgements				●

About these lessons

Lesson 1: Counting sounds

(m) Count reliably at least ten objects

Children tend to count objects by looking at them and touching them. This counting activity focuses on the senses of touch and hearing, giving children a broader sensory experience of matching counting numbers to objects.

Listen with sustained concentration

Classroom technique: Eyes closed, eyes open

In the introduction and plenary, children close their eyes while you drop counters into a pot. This means children must pay close attention, as they have only one chance to count the sounds of the counters dropping.

Improve learning and performance: reflect on learning

Children can begin to take responsibility for their own learning at an early age, as long as they are also helped to feel confident about their achievements. In this activity, gentle questioning encourages children to think about what they 'can do' and what they 'find hard'.

Lesson 2: Recognising patterns of spots

(m) Count reliably at least ten objects

When finding out how many of something there are, it is easier when you can recognise simple patterns: it is quicker to 'just see' five spots than to count them. In this activity, children say how many spots there are on a domino, making use of patterns of dots and 'counting on' to avoid having to count each spot.

Contribute to whole-class discussion

Classroom technique: Tell the class

In the plenary, children tell the class how they find the number of spots on a domino. You can support them by asking questions for them to answer. More confident and articulate children provide a model for others in the class to follow.

Work with others: show awareness and understanding of others' needs

Children working in pairs think about what their partners are finding hard or easy and adjust the challenges they set them accordingly.

Lesson 3: Odd and even numbers

(m) Recognise odd and even numbers

To work out whether a number is odd or even, children imagine the number as two rows of counters. Working with mental imagery is important in mathematics, and this game helps children develop the habit of visualisation.

Listen with sustained concentration

Classroom technique: Barrier game

Children turn away from each other or simply hold their cards so that their partner cannot see them. This means they must speak – and listen – to share the information on the cards.

Work with others: work cooperatively with others

Playing games – whether competitive and collaborative – encourages children to work cooperatively with others, as they take turns, speak and listen to each other and share decision making.

Lesson 4: Equal jumps on a number line

(m) Describe and extend number sequences

In this lesson, children make equal 'jumps' forwards and backwards on a number line. This prepares them for adding and subtracting numbers mentally and for working with multiplication. It also helps them develop a sense of the patterns in our number system.

Reach a common understanding with a partner

Classroom technique: Talking partners

Classroom techniques that structure the way in which children work in pairs have enormous value. However, children also need experience of working together in a more informal way. Here, children talk together about their joint task, take joint decisions and discuss how to share out the work.

Improve learning and performance: develop confidence in own judgements

The structure of the number line supports children as they work out which numbers they will land on when they draw jumps of two. Experiencing success with such activities helps children develop confidence in their own judgements.

Counting sounds

Classroom technique: Eyes closed, eyes open

Eyes closed, eyes open
Children need to learn not to cheat by peeping. However, as an additional insurance against peeping, turn the overhead projector off so that children cannot see the chosen number until they have all shown the number with their fingers.

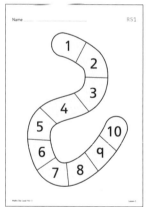

Introduction

Display RS1 to the class. Tell children to close their eyes. Put a cube on any number, then drop that many counters slowly and audibly into a tin or yogurt pot. Children count the sounds and hold up their fingers to show how many they counted.

Now tell children to open their eyes. Reveal the counters and check the true number by counting them with the class.

Leave the cube on the track and repeat the activity with another number.

m *This time, I'm choosing a number more than 5.*

Speaking *Does closing your eyes help you concentrate on the sounds?*

Personal *What goes on in your head when you are listening with your eyes closed?*

Pairs

Give each pair of children a copy of RS1, ten cubes and ten counters. One child from each pair closes their eyes and holds out their hands. The other child places a cube on a number on the track and then drops that many counters in their partner's hands, one at a time.

The first child says the number they think it is before opening their eyes and checking. If the child is correct, the pair leaves the cube on the track; otherwise, they remove it. They then swap roles and continue. Their aim is to get a cube on each number of the track.

m *Do you think you will manage to do all the numbers on the track?*

Support: Limit children to numbers up to 6.

Extend: Use the 1–20 track on RS2. Alternatively, one child drops 2p coins instead of counters. Their partner then works out the total value, which they check together, and puts a cube on the corresponding number.

Plenary

Thumbs up, thumbs down
Children put thumbs up to show when their numbers match and thumbs down to show when they don't.

Repeat the activity from the introduction: this time, children keep their eyes open. Tell them you may attempt to mislead them by dropping a number of counters different to the number you have marked. They must listen and work out whether or not you have dropped the number you indicated.

(m) *Did I drop in more or fewer counters than I said I would?*

(☺) *Do you think you're getting better at listening?*

Assessment for learning

Can the children

(m) Count up to 20 sounds – or beyond?

(☒) Listen to the counters being dropped into the pot without fidgeting or playing about?

(☺) Say one thing they did well in working on the task?

If not

(m) Assess whether the problem is with the counting or with imperfect listening. If you think it may be the former, do more work on accurate counting with a purpose.

(☒) Introduce a variety of short listening activities to give children practice in listening attentively.

(☺) Help children focus on one part of the activity at a time. Did they leave a space between each counter drop to help their partner distinguish sounds? Did they find the smaller numbers easy, but the higher numbers not so easy?

Recognising patterns of spots

Classroom technique: Tell the class

Learning objectives

(m) Maths
Count reliably at least ten objects

Speaking and listening
'Join in a discussion with the whole class'
Contribute to whole-class discussion

Personal skills
'Think about what other people need'
Work with others: show awareness and understanding of others' needs

(W) Words and phrases
number, zero, none, one, two, three... twenty, how many?, count, pattern, recognise, count on

(r) Resources
OHP dominoes or giant dominoes (optional)
for each pair:
set of double-six dominoes (or copies of RS3 and RS4)
set of double-nine dominoes (optional)

Two things to do
Emphasise to children that there are two things to do: to say the number and to say how they know. Their partner can remind them: "How do you know it's five?"

Counting on
With a whole domino, show one half first, say, "Put that number in your head", then show the other half and help children count on:

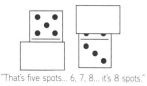

"That's five spots... 6, 7, 8... it's 8 spots."

Setting challenges
Ask children to notice how their partner is doing and to choose dominoes at a level of difficulty to suit.

Swapping over
The number of turns each child has is not prescribed. Encourage children to judge for themselves when they should swap over.

Introduction

Display one half of a domino to the class. Show it for about two seconds, then ask children to tell their partner how many spots they saw and whether they counted or 'just knew'.

Repeat this a few times, sometimes showing the whole domino.

(m) *Do I need to count every spot? How can I find the number without counting each one?*

(speaking) *Could you just see that it was four spots or did you count?*

Pairs

Give each pair of children a set of double-six dominoes or use RS3 and RS4. Child A shows a whole domino or half-domino to Child B, who says what the number is and explains how they know: did they need to count every spot or could they use the pattern of spots on the domino?

After a few more turns, the children swap roles.

(m) *I wonder if you counted a spot twice.*

(m) *How did you work out it was nine spots? Did you count on?*

(personal) *Is your partner finding this easy? How could you make it a bit harder for her?*

(personal) *Show me a domino you think might be a bit too hard for your partner. Shall we ask her if it is?*

(personal) *Why is it important to let the other person have a turn?*

Support: Use real dominoes and let children handle them in order to count the spots.

Extend: Use double-nine dominoes.

Plenary

Show dominoes one at a time. Each time, children tell you and the class whether they can recognise or work out the number of spots without counting each one.

Finally, show a domino pattern briefly before removing it. Children work in pairs to draw on a wipe board the pattern they think you showed and hold this up for you to see.

Do you recognise this pattern of spots? How many are there?

Jasmine said that when one side of the domino shows 1, she can work out the number of spots by adding 1 to the number on the other side. Does anyone else do that?

Tell the class
Note who has and who has not spoken. On another occasion, ask other children to talk about their work so that, eventually, everybody has had a turn. Support children with questions as necessary: "Could you just see the number of spots on one side from the pattern?"; "What did you do then? Show us what you did."; "You say you counted on. Show us how."

Assessment for learning

Can the children

Work out the number of spots on any domino by counting on?

Tell the class how they find out the number of spots on a domino?

Ask their partner if they want a 'harder' or an 'easier' domino?

If not

Do similar activities with fingers, recognising the number of fingers on one hand and counting on to find the total of fingers shown.

Ask a child who can describe what they do to tell the class their method and ask if anyone else did that. Support children with questioning to lead them through a description of their process.

Stop the children while you praise a pair who is doing this and recommend this way of working to the other children.

Odd and even numbers
Classroom technique: Barrier game

Learning objectives

(m) **Maths**
Recognise odd and even numbers

Speaking and listening
'Listen well'
Listen with sustained concentration

Personal skills
'Work well with others'
Work with others: work cooperatively with others

(W) **Words and phrases**
odd, even, equal, the same, number, zero, none, one, two, three... sixteen

(r) **Resources**
cards cut from RS5 (photocopy onto card)
counters

Odd and even
Talk about the numbers that can be arranged in pairs as 'even' and the numbers with one left over as 'odd'. Discuss how in an even number each counter has a partner, while in an odd number one of them doesn't.

Thumbs up, thumbs down
Children put thumbs up when they think a number is even and thumbs down when they think it is odd.

Methods
Children may picture the counters in their heads or may 'pretend' count them, as if they were in pairs. It may help children who are not using either of these methods to try one or both of them.

Introduction

Children work in pairs. Give each pair some counters. Write a number up to 10 on the board and ask children to predict whether that number of counters can be arranged in pairs without any left over.

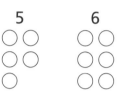

Pairs check with their own counters.

(m) *How do you know the counters will all go in pairs?*

(m) *How do you know there will be an odd one?*

Pairs

Give each pair of children a pile of cards cut from RS5. Pairs lay the cards face down on the table. Child A picks up the top card and reads out the number. Child B then says whether that number is odd or even. They check by putting counters in pairs.

Child A shows Child B the card and hands it to them if they were correct. Otherwise, they put the card to the bottom of the pile. After each card, the pairs swap roles.

(m) *What is an odd number? Why are 3 and 5 and 7 all odd?*

Can you repeat what Joe said about how he knows that 10 is even?

Both of you, tell me one way in which you think you are a good partner to work with.

 Listen well

Support: Use numbers up to 8 or 10. Give Child B counters so that they can model the number in pairs.

Extend: Use ordinary number cards showing higher numbers and provide Child A with a number line on which odd and even numbers are coloured differently.

Plenary

Write up a number. The class think about whether, with that number of children, everybody could have a partner, and explain why.

Repeat this process with other numbers.

Check one or two of the numbers by asking children to hold hands with their partners.

(m) *How do you know that 27 is odd?*

(☺) *Everyone, tell your partner one thing that is good about working with them.*

U&A Use reasoning and logic
During the main activity, talk to children about their methods for deciding whether a number is odd or even. Choose articulate children to explain these to others in the plenary: "I know 17 is odd because 7 is odd."

Assessment for learning

Can the children

(m) Recall or work out which numbers up to 10 (or 20) are odd and even?

(☒) Repeat what their partner has just said to them?

(☺) Help their partner to work out the answer without giving the game away?

If not

(m) Set up activities in which children work with sticks of cubes, trying to break them into two shorter sticks the same length. Provide practice in counting in jumps of two, starting from 1 and from 2. Ask children to colour alternate numbers on a number track.

(☒) Make a game of repeating words and phrases in a version of 'Simon Says' (children only repeat words and phrases that are preceded by the words 'Simon says...').

(☺) Comment in front of the class on those pairs who work well together and pinpoint what it is they are doing.

Equal jumps on a number line

Classroom technique: Talking partners

Tell your partner

Children take turns to tell their partner which number they think will be next.

Equal jumps

Help children count the jumps by 'bouncing' on the numbers in between and whispering their names, but landing properly on the even numbers: "One, *two*, three, *four*, five, *six*..."

U&A Work methodically to check solutions

Children look at the jumps they have drawn and check that they are all jumps of two – and that they started from 0. This will help develop the habit of checking work.

Introduction

Display RS6 and draw jumps of two along the number line from 0. Before each jump, ask the children to predict the next number you will land on.

(m) *What size of jumps am I drawing? Are they all the same size?*

(m) *What are the next two numbers I'll land on?*

Pairs

Give each pair of children a 0–20 or 0–30 number line (or a copy of RS6 or RS7) and ten counters. Each child circles five numbers on the line.

Children then draw jumps of two along the line from 0. For each circled number they land on, they take a counter from the pot.

Children record how many counters they have won and repeat the exercise, aiming to win more counters.

Tell George why you want to circle 10... George, do you agree?

Tell me what you are going to do first... And then?

Do you think you are getting good at guessing the numbers you'll land on?

Read out to me all the numbers you landed on. Can you say some of those without looking at the line?

Support: Children work with a number line to 10 or 20. An adult helper draws the jumps and asks the children to tell them which numbers to land on.

Extend: Children draw jumps of three, four or five. Alternatively, they start at 30 and jump back towards 0.

Plenary

Draw jumps of two starting from 1 on a 0–30 number line. Together with the children, say the numbers you land on.

Rub out the jumps you have drawn and ask each pair to write down three numbers they think you will land on if you repeat the jumps. Then draw the jumps again and tell the children to pat themselves on the back for each number they predicted correctly.

(m) *Will I land on the same numbers I landed on when I started at 0? Why not?*

(m) *Tell me some numbers I won't land on.*

(☺) *Shake hands with your partner and tell them they did really well in this lesson.*

Odd and even

Ask children what they noticed about the numbers they landed on when they jumped in twos starting from 0 (they are all even) and when they jumped in twos starting from 1 (they are all odd).

Assessment for learning

Can the children

(m) Describe what you are doing when you draw jumps of two on the number line?

(☒) Agree with their partner which numbers to circle on the line?

(☺) Confidently choose numbers that jumps of two from 0 or 1 will land on?

If not

(m) Draw equal jumps on a number line on a regular basis and sometimes talk about this yourself, as well as asking children to 'tell their partner' (p10) about the process. This helps familiarise children with the mathematical language needed.

(☒) Ask both children to explain which number(s) they want to circle, and why. Ask the child listening to repeat back the reason their partner has just given.

(☺) Make sure they are working at an appropriate level. Consider giving them work at a slightly easier level than they can handle for a while to give them experience of success and the confidence to tackle their work readily.

Self and peer assessment

Lesson 1: Counting sounds	I think	My partner thinks
(m) I can count up to 10 sounds.	🙂 ☹️	🙂 ☹️
👤 I listen carefully to the counters dropping so that I can hear them.	🙂 ☹️	🙂 ☹️

Lesson 2: Recognising patterns of spots	I think	My partner thinks
(m) I can say the number of spots on a half domino.	🙂 ☹️	🙂 ☹️
👤 I can tell the class about my counting.	🙂 ☹️	🙂 ☹️

Name _____

Lesson 3: Odd and even numbers	I think	My partner thinks
(m) I know which numbers up to 10 are odd and even.	🙂 😞	🙂 😞
(face icon) I can say what my partner has just said to me.	🙂 😞	🙂 😞

Lesson 4: Equal jumps on a number line	I think	My partner thinks
(m) I can draw jumps of two on a number line.	🙂 😞	🙂 😞
(face icon) I talk with my partner about our work.	🙂 😞	🙂 😞

Self and peer assessment

Place value and ordering

Learning objectives

	Lessons			
	5	6	7	8
Ⓜ Maths objectives				
count on or back from any number in steps of one	●			
order a set of familiar numbers		●		
know what each digit represents in a number to 20			●	
estimate a number of objects				●
Ⓢ Speaking and listening skills				
give accurate instructions	●			
listen and follow instructions accurately		●		
reach a common understanding with a partner			●	
talk about shared work with a partner				●
Ⓟ Personal skills				
improve learning and performance: critically evaluate own work	●			
organise work: check work		●		
work with others: work cooperatively with others			●	
work with others: show awareness and understanding of others' needs				●

About these lessons

Lesson 5: Number sequences

(m) Count on or back from any number in steps of one

This game helps children become familiar with the sequence made by counting in ones from a given number. It can be used at any level with any number sequence.

Give accurate instructions

Classroom technique: Eyes closed, eyes open

One child closes their eyes while the other makes a change to the number cards. The first child then opens their eyes and works out what has been changed before instructing their partner how to correct the deliberate error that has been introduced.

Improve learning and performance: critically evaluate own work

Children are encouraged to judge whether the work is 'hard' or 'easy' for them and to choose the range of numbers they work with accordingly.

Lesson 6: Ordering three numbers

(m) Order a set of familiar numbers

Children who find it easy to order consecutive numbers may find non-consecutive numbers more of a challenge. This activity teaches them a useful technique to help them order numbers.

Listen and follow instructions accurately

Classroom technique: One between two

Children share a pencil and a task. The child with the pencil writes down only what they are told to by their partner. This encourages the 'Instructor' to choose their words carefully and the 'Writer' to listen attentively to the instructions they are given.

Organise work: check work

Even young children can begin to acquire the habit of checking their own work before or when they finish it. In this activity, children are reminded to check their work.

Lesson 7: Place value to 20

(m) Know what each digit represents in a number to 20

Once children understand that the left-hand digit in a two-digit number represents the number of tens, numbers and calculations become easier. In this activity, children make and read numbers using place value ('arrow') cards to develop this understanding.

Reach a common understanding with a partner

Classroom technique: Barrier game

In situations where children are unable to see each other's work, they must concentrate on communicating verbally with clarity and precision to understand each other.

Work with others: work cooperatively with others

Pairs share an aim: to make numbers that match. With each number, they work together until they have achieved this aim.

Lesson 8: Estimating

(m) Estimate a number of objects

In real life, we use estimating perhaps as often as counting. Children can develop the skills of estimation in light-hearted activities such as this.

Talk about shared work with a partner

Classroom technique: Tell your partner

Children take turns to tell their partner their estimate about a number of counters. Simple tasks such as this encourage children to share information and ideas with a partner.

Work with others: show awareness and understanding of others' needs

Children briefly show their partner some objects. They need to think about the other person's needs, giving them enough time to make an estimate – but not enough time to count.

Number sequences

Classroom technique: Eyes closed, eyes open

Learning objectives

m Maths
Count on or back from any number in steps of one

Speaking and listening
'Give instructions'
Give accurate instructions

Personal skills
'Evaluate your own work'
Improve learning and performance: critically evaluate own work

W Words and phrases
number, zero, none, one, two... twenty, order, first, second, last, before, after, next, between

r Resources
large number cards (0–10 or 0–20)
'washing line' (optional)
for each pair:
blank cards
number lines (0–20 or 0–30)

Introduction

Display large digit cards from 0 to 20 (on a 'washing line' if available) and ask the children to help you put them in order.

Eyes closed
Children need to learn not to cheat by peeping. However, you can also ask children to bury their heads in their arms so you can see if any heads are raised.

Children close their eyes while you swap two numbers around. When they open them, they tell their partner which two numbers they think you have swapped.

One child then instructs you which cards to change to restore order.

m *Read out those numbers. Do they sound right?*

What must I do to put those numbers right?

Pairs

Give each pair of children about ten blank cards and a starting number. Pairs write a number on each card, making a sequence from that starting number.

Evaluating their own work
You could also encourage children to choose their own starting number. After a while, stop the class for a mini-plenary. Children judge whether they are finding the work hard or easy and reduce or extend their number sequence accordingly. Some children may be ready to work with a new sequence entirely.

Giving instructions
Emphasise to children that the child who closed their eyes must tell their partner which cards to move – they must *not* touch the cards themselves or point. You could insist on hands on laps or behind backs.

They then take turns to close their eyes while their partner swaps round two cards. When they open their eyes, they instruct their partner which cards to change to restore order.

m *Can you count on from 4 instead of starting at 1?*

m *Show me that number on the number line. What comes next?*

Do you understand what Stacy is saying? If not, what could you say to her?

Are these numbers too easy for you? Or just right?

What numbers will you try working with next? Show me on the number line.

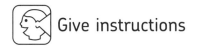
Support: Provide ready-numbered cards. Limit the activity to numbers up to 10. Provide a number line for children to use as a check.

Extend: Give children more cards. Encourage children to work with sequences that cross a tens boundary – for example, 14 to 27 or 23 to 39.

Plenary

Ask a child to hold up two of their digit cards and to sread one of the numbers. The class count from that number until they reach the second number – so they may need to count backwards or forwards.

Repeat this process with other children and their sequences.

ⓜ *Sian's numbers are 19 and 8. If you start at 19, What number will you say next?*

Assessment for learning

Can the children

ⓜ Count on in ones from 5? From 12? From 32?

🗣 Make it clear to their partner, without pointing, which two cards to swap round?

🙂 Choose a sequence of numbers to work with that challenges them appropriately?

If not

ⓜ Practise counting round in a group, supported by a number line.

🗣 Change the activity so that one child hides a card and closes up the gap in the sequence before the other child identifies the missing number.

🙂 Encourage children to try something harder if they choose work that is too easy and reassure them that they will be successful. If they choose work that is too hard, tell them you are impressed that they want challenging work, but explain that it is best to work consistently, moving from easier to harder levels.

Ordering three numbers

Classroom technique: One between two

One between two
Explain that the child who does not have a pencil decides what to write and tells their partner – who has the pencil – where to write it. If the instructions are not clear, the child with the pencil asks their partner to repeat them or ask questions to clarify what is meant.

Introduction

Display RS8 to the class.

Ask for a volunteer to help you model 'One between two' (p8). Roll a 0–9 dice three times and ask your partner to write the numbers in the first set of boxes on RS8 (under the 'Dice rolls' heading).

Start counting on from 1. When you come to a number that your partner has previously written on RS8, instruct them to write that number in the first box on the right (under the 'In order' heading). Continue reciting the sequence of numbers like this until all three numbers are written in.

Swap roles and work together to order another set of numbers in the same way.

Then check your work by looking at the numbers with a number line or track.

(m) *How can I work out what number comes next after 5?*

(s) *Are there any other numbers that belong between 2 and 7?*

(p) *Help me check if the numbers are in order.*

Pairs

Give each pair of children their own copy of RS8. Pairs repeat the process you have just demonstrated.

After two turns each at ordering three numbers, the pairs move on to ordering four numbers.

(m) *Which comes first, 9 or 12? How do you know?*

(s) *Where does Kenny want you to write the 7?*

(s) *Tell me what Anna just said to you.*

(p) *How will you check your work? Why is checking work a good idea?*

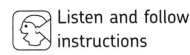
Support: Use a 1–6 dice. Give children a number line to remind them of the number order. Ask an adult helper to work with anyone experiencing difficulties.

Extend: Use a 1–20 dice – or even 0–100 cards – and use RS9 to work on ordering five numbers.

Plenary

Each child chooses a number from 1 to 30 and writes it on a piece of paper. Children work in groups of four, combining their numbers and putting them in order. Two such groups then work as a group of eight and order all eight numbers.

At the end, each group of eight read out their ordered numbers.

How can you make sure that your numbers are in order?

Checking their work
Ask children to check their work using the reciting technique if they find ordering the numbers too difficult.

Assessment for learning

Can the children

Put in order three or more non-consecutive numbers below 10? Below 20? Below 50?

Repeat back what their partner has just asked them to do?

Spot any errors they have made?

If not

Check that children can say the number sequence. Continue with the initial technique of reciting the number sequence until you come to one of the numbers to be ordered.

Ask the partner to say a bit at a time and encourage the child to repeat back just these words: "write the 5" – "write the 5"; "in the second box" – "in the second box".

Do some work with the class on spotting deliberate errors. Ask pairs to check each other's work.

Place value to 20
Classroom technique: Barrier game

Saying which cards to use
For example: "We made 17. We used the 10 card and the 7 card."

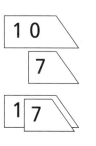

Base board
The purpose of this is to encourage pairs to share the place where they put the number they have made. This avoids a situation in which one child is holding on to the cards.

Barriers
Children sit back to back, or Child A simply keeps their number hidden.

Supporting each other
If appropriate, Child A can support their partner by telling them which cards to use.

Introduction

Children work in pairs. Give each pair a set of place value arrow cards for numbers to 30 and a base board (or copy of RS10) on which to place their number.

Use demonstration place value arrow cards to make various numbers up to 30 or above. Each time you do this, pairs make the identical number using their cards and place it on their base board.

Pairs say which cards they used to make the number.

m *How many cards do I need to make the number 24?*

Speaking *Tell your partner how to change the 24 to 26.*

Pairs

Give each pair of children a copy of RS11. Child A circles a number on the number grid on RS11. They read this out to Child B, who then makes the number on the base board, using the place value arrow cards.

Children both look together to check if their numbers are the same. If they are not the same, they decide together how to change the cards.

Children then swap roles and repeat the process.

m *Whisper to me which number you've circled. Which bit of that is the tens digit?*

m *That card there shows 10. When you cover the zero with 5, what is the new number?*

Speaking *Tell Shannon how many cards she needs. Does she need the 10 card?*

Personal *How do you feel when somebody shares the cards and doesn't hoard them? Does it feel good?*

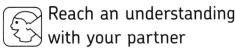
Support: Use numbers from 1 to 20.

Extend: Use cards to 50 and the 1–50 grid on RS12.

Plenary

Pairs share their place value arrow cards and work together to hold up numbers one (or ten) more (or less) than a certain number: "the number one more than 8"; "ten less than 25".

 Show me 15. Now show me the number that is ten more than 15. Which card do you need to change?

Talk to your partner about how you are sharing the work. Are you happy with it or do you want to change anything?

Assessment for learning

Can the children	**If not**
Make numbers to 20 or 30 accurately, with the digits in the correct order?	Do some activities with the class (using demonstration place value arrow cards) in which the children make and read numbers.
Agree which place value cards are needed to make the number circled on the grid?	Ask both partners to say which cards they think are needed, and why.
Support their partner if they get into difficulties or make an error?	Talk about the value of helping others – but not doing their work for them. Model supportive behaviour yourself when working with the child in question.

Estimating
Classroom technique: Tell your partner

Learning objectives

(m) Maths

Estimate a number of objects

Speaking and listening

'Talk about your work with your partner'
Talk about shared work with a partner

Personal skills

'Think about what other people need'
Work with others: show awareness and understanding of others' needs

(w) Words and phrases

number, zero, none, one, two, three... ten, twenty, guess, estimate, nearly, roughly, more, less, fewer

(r) Resources

display copy of RS13 (optional)
display number line
for each pair:
countable objects
sheet of paper
copy of RS14
0–30 number line (optional)

Pats on backs
The actual number of same-sized stars is 8, so estimates of 7, 8 or 9 get a pat on the back.

How much time?
Be specific that this is one of their tasks – to think about how long their partner needs to estimate the number of objects without actually counting.

Introduction

Briefly display one section of RS13 to the class, then cover it up.

Children 'tell their partner' (p10) how many same-sized objects they think you showed them. They can also hold up their fingers to show you.

Reveal the objects again and count them with the class. Tell children to pat themselves on the back if they were correct or if they guessed a number next to the actual number. Use a number line to check this.

Talk about the difference between guessing (where there is little or no evidence) and estimating (where there is some evidence available). Discuss techniques you or the children use when estimating.

Repeat this with the other sections of RS13.

(m) *There are two stars here and more here, so the number must be more than 2.*

(m) *That doesn't look as many as the last time, so it must be less than 8.*

(m) *It was 12. Which numbers are next to 12?*

(·) *Tell your partner whether you think there were more than five circles.*

Pairs

Give each pair of children a pencil, a pot of about 20 countable objects, a copy of RS14 and a sheet of paper. Child A secretly takes some objects and spreads them on the table. They briefly show them to Child B before covering them with the sheet of paper.

Child B says out loud their estimate of the number of objects, and Child A records this on the first number line on RS14.

The children then count the objects together and record the actual result on the same line before swapping roles and repeating the process.

How is your estimating? Let's look at the number lines and see how close your estimates are to the actual number.

Tell your partner why you think there were about eight different-size shapes.

Tell your partner if they are giving you enough time to make an estimate, or too much time, or whether it is just right.

Do you need to make the estimating a bit easier? How could you do that?

Support: Use a pot of ten objects.

Extend: Use 30 counters and a 0–30 number line.

Plenary

Outline the three objectives of this lesson and talk about how practising these skills will help the children get better at them.

Children then turn to their partners and talk briefly about which of the objectives they think they have achieved best in this lesson.

Tell your partner whether you think you have got better at estimating how many objects there are.

Did you do well at talking about your work with your partner?

Assessment for learning

Can the children

Make a reasonable estimate of a number of objects?

Speak clearly and look at their partner when telling them their estimate?

Show consideration for their partner by displaying the objects for a suitable amount of time?

If not

Check whether they can estimate smaller numbers and model estimation in class: "That must be about six times we've read this story"; "We can fit about another three people round this table"; "I estimate there are about 20 children on this carpet. Let's count and see."

Make a game in which children whisper numbers, shout them and speak them clearly while looking at you or a partner.

Read or tell stories and ask children to talk about how it might feel to be one of the characters in the story.

Name _____

Self and peer assessment

Lesson 5: Number sequences	I think	My partner thinks
(m) I can count on in ones from ☐	🙂 ☹️	🙂 ☹️
👥 I tell my partner which cards to swap round.	🙂 ☹️	🙂 ☹️

Lesson 6: Ordering three numbers	I think	My partner thinks
(m) I can put in order three numbers up to 10. I can put in order three numbers up to 20. I can put in order three numbers up to 50.	🙂 ☹️	🙂 ☹️
👥 I can repeat back what my partner has just said to me.	🙂 ☹️	🙂 ☹️

Name _____

Lesson 7: Place value to 20	I think	My partner thinks
(m) I can make numbers up to 20 with arrow cards.	🙂 ☹️	🙂 ☹️
I talk about numbers with my partner, and we can match them.	🙂 ☹️	🙂 ☹️

Lesson 8: Estimating	I think	My partner thinks
(m) I can estimate a number of things.	🙂 ☹️	🙂 ☹️
I speak clearly when I tell my partner about my estimate.	🙂 ☹️	🙂 ☹️

Self and peer assessment

Calculations

Learning objectives

	Lessons			
	9	10	11	12
Maths objectives				
know by heart pairs of numbers that total 10	●			
know or derive addition facts to 12		●		
understand that more than two numbers can be added together			●	
find the total value of two or more coins				●
Speaking and listening skills				
speak confidently in front of the class	●			
listen and follow instructions accurately		●		
reach a common understanding with a partner			●	
contribute to whole-class discussion				●
Personal skills				
work with others: work cooperatively with others	●			
organise work: organise findings		●		
organise work: identify stages in the process of fulfilling a task			●	
organise work: check work				●

About these lessons

Lesson 9: Number bonds to 10

(m) Know by heart pairs of numbers that total 10

Children who know the pairs of numbers that make 10 can use these to work out other number combinations, while those who don't can be disadvantaged throughout their lives. Games such as this encourage children to use mental modelling to work out – and remember – these crucial facts.

Speak confidently in front of the class

Classroom technique: Tell the class

Children talk to the class about how they solve a simple number problem. The teacher can support them by asking questions for them to answer. More confident and articulate children provide a model for others in the class to follow.

Work with others: work cooperatively with others

Children take turns to do the 'magic trick' for the benefit of their partner, and then pairs work together to prepare one more trick to show the class. At all these different stages, cooperative working is essential.

Lesson 10: Adding two numbers

(m) Know or derive addition facts to 12

Activities such as this, in which children roll two dice and use a number line to add the numbers, help children slowly learn facts by heart. This type of activity also gives them a bank of mental imagery that will eventually allow them to abandon the actual line and work mentally.

Listen and follow instructions accurately

Classroom technique: One between two

Children share a pencil and a task. The child with the pencil writes down only what they are told to by their partner. This means that both children must concentrate on communication, giving instructions and following them.

Organise work: organise findings

The structured resource sheet used in this activity helps children record their work in a consistent and comprehensible way. This provides a model for organising and recording their findings when working more independently.

Lesson 11: Adding three numbers

(m) Understand that more than two numbers can be added together

It is all too easy to give children the impression that addition is all about combining two numbers. They need to know that, for example, three numbers can be added – as they discover in this activity.

Reach a common understanding with a partner

Classroom technique: Talking partners

Classroom techniques that structure the way in which children work together in pairs have enormous value. However, children also need to experience working together in a more informal way. In this activity, children talk together about their joint task and share responsibility for the accuracy of their work.

Organise work: identify stages in the process of fulfilling a task

Pairs are encouraged to carry out the task and to organise and record their findings in their own way. A structured resource sheet is provided to support less confident children, assisting them in recording their work consistently and clearly.

Lesson 12: Coin totals

(m) Find the total value of two or more coins

Children are supported in finding the totals of coins by working with a partner. Pairs work together using a number line to add the coin values.

Contribute to whole-class discussion

Classroom technique: Heads or tails

Children share a task, knowing that either one of them may be asked to talk about it in a class discussion. This knowledge gives them an incentive to take an active part and not to rely on their partner to do the work.

Organise work: check work

Children are asked to check their work on completion. As the class are all working on the same problem at the same time, there is an opportunity for you to remind them of this with each calculation, which will help make checking habitual.

Number bonds to 10

Classroom technique: Tell the class

Learning objectives

m Maths
Know by heart pairs of numbers that total 10

Speaking and listening
'Speak in front of the class'
Speak confidently in front of the class

Personal skills
'Work well with others'
Work with others: work cooperatively with others

W Words and phrases
add, more, plus, make, total, altogether, how many more to make, how much more/less, equals, whole

r Resources
linking cubes for each pair:
copy of RS15

Working out the complement
Show how to do this with your fingers: "I can see four cubes, so here are four fingers. I know the whole stick had ten cubes in it, so I counted these other fingers... and there are six. Four fingers and six more makes ten."

U&A Using mental imagery
Encourage children to make simple calculations 'in their head'. Manipulating mental images is an important aspect of mathematical thinking at all levels.

Introduction

Tell the children that you are going to teach them a 'magic trick' – or one that can look like magic.

Give a child a stick of ten cubes, but don't say how many there are. The child breaks the stick of cubes in two and hides one part (perhaps under a book or behind their back). They then hold up the remaining part.

Say that you are going to tell them how many cubes are hidden. Then count the cubes being held up and announce the hidden number.

on show · behind the child's back

The child reveals the hidden part, and everybody counts the cubes in it as a check.

Finally, explain how you did the 'trick': by counting the cubes in the child's hand and working out how many must be added to make 10.

Invite two confident children to repeat the 'trick' in front of the class.

m *I can see that Joss has six cubes. So I make that up to 10: 7, 8, 9, 10* [counting on your fingers]. *Joss, are you hiding four cubes?*

Show and tell us how you worked that out.

Pairs

Give each pair of children a copy of RS15. Pairs repeat the same activity, taking turns to break a ten-stick and to hide part of it. The partner works out the number of hidden cubes. Pairs record their work on RS15.

Pairs then prepare one such 'trick' to show the class, using two different colours of cubes so that they will know where to make the break.

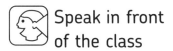
(m) *How do you know that Salim doesn't have six cubes hidden?*

(☺) *What does 'working cooperatively' mean?*

(☺) *Why is it important to work cooperatively with your partner?*

Support: Work with a stick of six cubes. Alternatively, show children how to use the template of a ten-stick on RS 15 to lay one piece on and work out the missing part.

Extend: If children are sure of the number bonds to 10, give them an eleven-stick to work with. Ask them how their knowledge of the bonds to 10 can help them with the bonds to 11.

Plenary

Pairs demonstrate one of their 'tricks' to the rest of the class. The child who breaks the stick talks about what they are doing as they do it: "I've got a ten-stick. Now I'm breaking it in two pieces…" The other child explains about how they work out the size of the missing part.

(m) *James and Phil made an eleven-stick. Would your method work with that?*

(☺) *Pippa, I saw you nodding your head when you worked out the hidden number. Were you counting? Tell us what you did.*

Tell the class

Ask half the children to talk in this way and note who has spoken. On another occasion, ask the rest of the children to talk about another piece of work. Support children as necessary with questions: "Did you just know the number that goes with 7 to make 10?"; "Did you use your fingers? Show us what you did"; "You say you did it in your head. Tell us how."

Another 'magic trick'

Show how you can use one fact to work out another – for example, if seven cubes were showing when three were hidden, then you know that when three are showing, seven must be hidden.

Assessment for learning

Can the children

(m) Work out some missing numbers either mentally or with their fingers?

(☺) Talk to the class about the 'magic trick' they have prepared?

(☺) Readily take turns with their partner?

If not

(m) Check which number bonds below 10 they can work out and help them build on these: "You know that 3 and 3 makes 6, so what does 3 and 4 make?" Build up smaller bonds before coming back to bonds to 10.

(☺) Prompt children with questions to which they can simply answer 'Yes' or 'No' or even nod or shake their head. Offer praise for being brave enough to stand up in front of the class.

(☺) Ask a pair who works cooperatively to tell the class why it is a good idea, and how they feel about working in this way.

Adding two numbers

Classroom technique: One between two

Learning objectives

m Maths
Know or derive addition facts to 12

Speaking and listening
'Listen and follow instructions'
Listen and follow instructions accurately

Personal skills
'Organise your results'
Organise work: organise findings

W Words and phrases
add, plus, make, more, extra, total, altogether, equals, record, write, column, symbol

r Resources
0–30 number lines and felt-tip pens (optional)
display copy of RS16 (optional)
for each pair:
copy of RS16 and RS17

Recording
Use the board or a display copy of RS16 to record the dice numbers and addition equation.

Put the larger number first
Remind children that this makes the calculation easier, as there are fewer steps to make on the line.

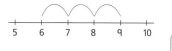

One between two
Tell pairs that the child with the pencil is responsible for carrying out their partner's instructions. If these instructions are not clear, the child asks their partner to repeat them or asks questions to clarify what they mean.

Introduction

Revise with the class how to use a number line for addition. Roll two dice, then instruct a volunteer what to draw on the number line to add the two numbers. Also tell the child how to record this as an equation.

Swap roles with your volunteer and invite the child to tell you how to add the next two dice numbers.

m *Why is it easier to put the bigger number first?*

What do you want me to do first? How many steps must I draw?

I've written the two dice numbers. What do I write now? And where do I write it?

Pairs

Give each pair of children two dice and a copy of RS16. Pairs continue the process in the same way, swapping roles after every addition.

Once they have finished RS16, children work with RS17, using the number lines to do the additions, but recording their findings in their own way.

m *You've rolled a 6 and a 2. How will you use the number line to add those numbers?*

Do you agree with what Ben just asked you to do?

Has Maria written that 2 the right way round? Can you tell her how to write it?

Where do those dice numbers go? And what goes in this column?

How can you organise your recording so that I will understand it?

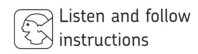
Support: Adapt the dice to show only numbers up to 3 or 4.

Extend: Children organise their own recording, without using RS16. Give pairs two 0–9 or 1–12 dice and a 0–30 number line. Remind them to work mentally whenever they can.

Plenary

Review some of the skills ('mathematical', 'speaking and listening' and 'personal') that children have been practising.

Children choose one skill they think they used well in this lesson and one skill they think they need to practise more. They tell their partner about these and find out if the partner agrees.

Tell your partner one time today when you listened well.

Tell your partner whether you think you could add two dice numbers on a number line, then record the addition all on your own.

Skills list

For example:

– listening to a partner

– being patient

– helping a partner

– using a number line to add two numbers

– using mental addition

– organising their recording

– checking their work

Assessment for learning

Can the children

Add two small numbers either mentally or on a number line?

Record their work as instructed by their partner?

Write the numbers and equations so that they are comprehensible to others?

If not

Check whether children can use their fingers and count on from one number. If not, focus on the technique of counting on. Play games and set up activities based on adding small numbers to give children confidence and to reassure them that there are some bonds they do know.

Introduce mathematical vocabulary, and use the symbols '+' and '=' to familiarise children with them in a variety of contexts.

Practise making simple tables with the class. Invite children to add data to, and then interpret, the table on the board.

Adding three numbers
Classroom technique: Talking partners

Learning objectives

m Maths
Understand that more than two numbers can be added together

Speaking and listening
'Reach an understanding with your partner'
Reach a common understanding with a partner

Personal skills
'Keep track of where you are'
Organise work: identifying stages in the process of fulfilling a task

W Words and phrases
add, plus, make, enough, at least, total, altogether, is the same as, equals, record, write, column

r Resources
for each pair:
plastic or real beans
copy of RS18 (optional)
0–5 dice or paper clip spinner (either adapt the 1–6 dice or use the spinner on RS19)
1–3, 1–4 or 1–6 dice (optional)

Modelling with beans
Act out the story, with a boy, a girl and a 'cat' and some large beans. Make sure everyone can see the beans and count them together. Agree whether they have ten beans or more.

Talking partners
The matter for discussion in this lesson is as much the way of working and recording as the actual adding. Ask children to talk about how to share the work fairly and to make sure that they both agree about the calculations and recordings they do.

Introduction

Tell the class a story about a boy, a girl and a cat who need at least ten magic beans to get the witch to release them from the gingerbread house. A giant gives each of the three a number of beans according to the roll of a dice.

Roll the 0–5 dice to find out how many beans the boy will get, again to find out how many beans the girl will get and once more to see how many their cat will get.

Discuss with the class how to total the three sets of beans and decide whether they have enough.

m *Could just the boy and girl have enough beans if there was no cat? What dice numbers would they need?*

m *Tell us why you think they don't have enough beans.*

Pairs

Pairs of children work together to generate a number using a dice and count out that many beans. They do this three times (once each for the boy, the girl and the cat) and record their numbers and total. They also need to decide whether or not there are enough beans to get the witch to release them.

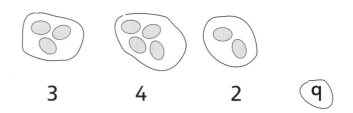

3   4   2   9

Pairs do several calculations in this way.

m *The boy and girl's numbers are the same as last time, but swapped around. What will their total be then? How do you know?*

m *How do you put out zero beans? How will you show that the boy had zero beans?*

☺ *Show Jack how you add those numbers.*

Reach an understanding with your partner

What do you need to do next?

How can you check that total is correct?

Tell me about this calculation. Is it complete?

Support: Use 1–3 or 1–4 dice. Give children a copy of RS18 to make their records.

Extend: Use 1–6 dice. Encourage children to work mentally or with a number line, rather than using beans.

Plenary

Thumbs up, thumbs down
This allows you to make a quick assessment of how well the children are coping with this kind of calculation.

Ask one of the pairs to come to the front of the class. One of the children chooses one of their calculations and reads out all three numbers of beans. As the child reads, both children hold up their fingers to represent these numbers. Then the rest of the class show 'thumbs up' or 'thumbs down' to indicate whether they think the total is enough (ten or more).

As a class, carry out the calculation by counting the pair's fingers. Discuss ways to simplify the calculation, such as looking for a pair of numbers that totals 10 or starting with the largest number and counting on.

Is it possible to add three numbers using one person's fingers? When can't you?

Have we got the answer yet? What do you think it is?

Assessment for learning

Can the children

Add three small numbers accurately?

Talk about their shared work, even if their partner did the recording?

Say what they have done and what they still need to do?

If not

Help children touch each bean if they have poor counting skills and move it aside as they say the number.

Ask the child or pair to carry out a calculation, while you watch and describe what they are doing as they do it.

Help children by asking questions such as: "What is the word here? And what does the number here tell you? Yes, how many beans the girl has. And this?"

Coin totals

Classroom technique: Heads or tails

Introduction

Display some 1p, 2p, 5p and 10p coins to the class. Revise with children how to find the total of two or three coins, using the strategy of putting the highest value coin first and counting on. Use a number line or fingers as appropriate.

(m) *I'm starting with the 10p coin and adding on 2p. Show us what to do on the number line.*

I've got a 5p and a 2p coin. To add their values, which coin shall I start with? Why should I start with the 5p coin?

Tell us how you know that 5p and 1p doesn't make 7p.

Heads or tails
Tell children that you may call on either partner to speak to the class, so both must take equal responsibility for understanding their work.

Checking work
Make a point of asking children to check each time that their coins do total the correct value and to adjust their selection if they don't.

Pairs

Give each pair of children a pile of 1p, 2p, 5p and 10p coins and a copy of RS6 (see Lesson 4).

In secret, choose two or three coins, calculate their total and put them in your pocket. Tell the class the total and ask pairs to put together a set of coins that totals your value.

Two or three pairs come to the front of the class and give their suggestions as to what your coins might be. Record these suggestions.

Vary the game
To improve children's chances of guessing your coins correctly, give them a clue. For example, tell them how many coins you have, or the value of one or two of the coins. Be aware, though, that using this information requires more sophisticated thinking than just telling them the total.

> **Total 7p**
> 5p 2p
> 2p 2p 2p 1p
> 1p 1p 1p 1p 1p 1p 1p

Discuss with the class and individual pairs how to add the sets of coin values with fingers, on the number line or mentally and then let them do this. Finally, reveal your coins. Emphasise that all guesses are equally valid (as long as the totals are correct) and that there are no winners or losers. Continue with another set of coins.

Who can say how to add those two coin values using fingers? How shall we do it?

Can you explain why you think the coins might be a 2p and a 5p piece?

Do you understand Megan's method of adding the coins?

How can you check that you've added those coins correctly?

Support: Children speak to the class only in instances where the total is of low value and can be made from 1p, 2p and 5p coins.

Extend: Expect children to make use of clues as suggested in 'Vary the game'.

Plenary

Display RS20 to the class and make one last coin selection, this time totalling 10p.

Work together as a class to find six different ways to make 10p and record each set of coins in one of the 'pockets' on RS20.

Is that set of coins different from the ones we have recorded so far?

Are there any other ways to make 10p? What could we try?

Why is it important to check our work?

Assessment for learning

Can the children

Find a reliable way to add two or more coin values?

Offer a suggestion or contribution that is appropriate to the matter in hand?

Check their work and spot and rectify errors?

If not

Set up some pairs in a 'Peer tutoring' (see p8) situation, where one confident child teaches another child how to solve these problems.

Support children by asking questions such as: "What is the first thing you do? And how does that help you?"

Give the children a few calculations that include one or two deliberate errors and ask them to spot them. Have a 'checking week' during which children regularly check and correct their own or their partner's work.

Self and peer assessment

Lesson 9: Number bonds to 10	I think	My partner thinks
(m) I can work out how many cubes my partner is hiding.	☺ ☹	☺ ☹
I can tell the class about one of my 'magic tricks'.	☺ ☹	☺ ☹

Lesson 10: Adding two numbers	I think	My partner thinks
(m) I can add two dice numbers.	☺ ☹	☺ ☹
I listen carefully to what my partner tells me.	☺ ☹	☺ ☹

Name _____

Self and peer assessment

Lesson 11: Adding three numbers	I think	My partner thinks
(m) I can add three small numbers.	🙂 ☹️	🙂 ☹️
🗣 I can explain what the recordings I made mean.	🙂 ☹️	🙂 ☹️

Lesson 12: Coin totals	I think	My partner thinks
(m) I can add two or three coin values.	🙂 ☹️	🙂 ☹️
🗣 I can say something in a class discussion.	🙂 ☹️	🙂 ☹️

Handling data

Learning objectives

	Lessons			
	13	**14**	**15**	**16**
ⓜ Maths objectives				
solve a given problem by sorting	●			
collect data and organise a table		●		
make a table: sort and organise information			●	
interpret data				●
Speaking and listening skills				
listen with sustained concentration	●			
contribute to whole-class discussion		●		
contribute to small-group discussion			●	
reach a common understanding with a partner				●
Personal skills				
improve learning and performance: take pride in work	●			
improve learning and performance: reflect on learning		●		
work with others: discuss and agree ways of working			●	
improve learning and performance: develop confidence in own judgements				●

About these lessons

Lesson 13: Sorting numerals

(m) Solve a given problem by sorting

In this activity, children sort numerals according to one criterion. This simple sorting gives children experience of asking a question about each item before consigning it to a given set.

Listen with sustained concentration

Classroom technique: One between two

Children share a pencil and a task. The child with the pencil writes down only what they are told to by their partner. This encourages the child with the pencil to listen carefully.

Improve learning and performance: take pride in work

All school activities provide children with opportunities to take pride in their work. Every now and then, it is a good idea to make this a particular focus of a lesson and help children think about which qualities of their work are worthy of pride.

Lesson 14: Organising a table 1

(m) Collect data and organise a table

In this activity, children vote for which story they would like to hear. Voting, then recording the votes in a table, is a way of organising information about opinions. Children will want to make sense of the table as the subject matter is important to them.

Contribute to whole-class discussion

Classroom technique: Talking stick

To ensure that children listen to each other and do not 'talk over' each other, use a 'talking stick'. The child who holds the stick has a chance to speak about their ideas – and the right to be listened to without interruption.

Improve learning and performance: reflect on learning

Children are encouraged to think about the process of voting by the teacher's careful questioning.

Lesson 15: Organising a table 2

(m) Make a table: sort and organise information

Children make a table showing the numbers of objects in their pot. This is a development from the previous lesson in which the teacher makes a table using the class's data.

Contribute to small-group discussion

Classroom technique: Talking partners

Classroom techniques that structure the way in which children work in pairs have enormous value. However, children also need to practise working together in a more informal way. In this activity, a small group talks together about their joint task, deciding for themselves how to fulfil it and discussing the teacher's questions together.

Work with others: discuss and agree ways of working

In this activity, children have to make decisions as a group, such as who will count which set and how to organise filling in the table. This will help them think about how to share out work fairly and involve others in decision making.

Lesson 16: Interpreting data

(m) Interpret data

Children sort themselves according to their favourite sporting activity or game. They then discuss whether the statements you make about the data are true or false.

Reach a common understanding with a partner

Classroom technique: Devil's advocate

Establish with the class that the statements you make about the data may or may not be true. Pairs discuss them to reach agreement about whether or not each statement is true.

Improve learning and performance: develop confidence in own judgements

Children have to make their own judgements about the data to decide on the truth of your statements. Sometimes, they will need to disagree with what has been said (by you or a puppet). This helps them develop the idea that they can make their own valid judgements and abide by them.

Sorting numerals

Classroom technique: One between two

Learning objectives

(m) Maths
Solve a given problem by sorting

Speaking and listening
'Listen well'
Listen with sustained concentration

Personal skills
'Take pride in your work'
Improve learning and performance: take pride in work

(w) Words and phrases
numeral, set, sort, group, all the same, curved, corners, straight

(r) Resources
wooden or plastic numerals (optional)
for each pair:
copies of RS21
mirrors (optional)

Properties of numeral shapes
Include the descriptions on RS21: straight lines, curves, whether you can write the number without taking your pencil off the paper, corners, whether it looks the same upside down or in a mirror.

One pencil between two
Children take turns to choose a numeral and instruct their partner which set to write it in. They then cross out that numeral on the number line at the top of the sheet.

Making decisions
Children may (justifiably) not be sure whether, for example, 2 or 9 has corners. Encourage them to discuss this and make these decisions for themselves. Reassure them, if necessary, that there is no clear correct answer – and that their opinion is valid.

Introduction

Display numerals from 0 to 9 (written or printed, or in wood or plastic) to the class. Talk about their shapes with the children.

(m) *Shall I see if I can write the 9 without taking my pen off the board? Put your thumb up if you think I can.*

(m) *Tell your partner which numbers you think have straight lines. Which ones have **only** straight lines and no curves?*

Pairs

Give each pair of children a copy of RS21. Pairs choose one of the questions and circle it. They then write out the numerals in the sets below accordingly.

When children have completed this, they check and correct their work before sorting the numerals a different way on another copy of RS21.

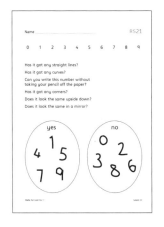

(m) *What is the same about all these numerals? How are they different to the ones in this set?*

(m) *Did you both agree about every numeral? Did you sometimes have to talk about which set to put one in?*

(s) *Which set did Sarah say to put the 3 in? Why did she say it goes there?*

 Listen well

Taking pride in work
Ask children what makes them feel good about a piece of work and proud of what they have achieved.

Inventing criteria
Help them formulate their ideas as a question with a clear 'Yes' or 'No' answer – for example: "Can you make the letter with straight lines?"

Are you pleased that you have sorted those numerals in different ways?

Which of those sheets are you most pleased with?

Support: Children sort wooden or plastic numerals onto two sheets of paper.

Extend: Children choose their own criteria to sort capital letter shapes.

Plenary

Children talk about any ideas or problems they have come up with during the activity.

Does 8 look the same in a mirror? How can we check?

You said something interesting to me about upside down numbers. Tell the class what you said.

Assessment for learning

Can the children

Sort the numerals consistently by one chosen criterion?

Repeat what their partner has just said?

Identify a piece of completed work that they are proud of and say what they like about it?

If not

Do some activities in which you sort children by one criterion ('wearing shorts or not', 'has a sister or not', and so on) and discuss the process as you do so.

Make a game of saying a simple word or sentence for the children to repeat.

Check whether the work was too easy or difficult or whether children feel their recording is untidy. Tell children which piece of their work you are most pleased with, and why.

Organising a table 1
Classroom technique: Talking stick

Learning objectives

m Maths
Collect data and organise a table

Speaking and listening
'Join in a discussion with the whole class'
Contribute to whole-class discussion

Personal skills
'Think about what you have learned'
Improve learning and performance: reflect on learning

W Words and phrases
count, set, sort, vote, group, table, chart, collect, most, least, favourite

r Resources
four or five pots, each labelled with the title of a different story from the class library
linking cubes
'talking stick' (see p10)

Reflecting on what they have learned
Help children think about the voting process by focusing on what it can be used for – and how it might be abused.

Introduction

Show the children the four or five pots that you have labelled with titles of stories from the class library. Read out the labels and explain that the children are going to choose which story you will read today. Say that to make sure that the decision is fair, they are all going to vote.

Give each child one of the linking cubes. Children put their cube in the pot labelled with the story they choose.

Tell us whether you think this is a fair way of choosing which story to hear.

What might make it unfair?

What other things could we use voting for? When might it be useful?

Whole class

Show the pots around and talk to the class about how you cannot be sure which one has most in it without counting.

Tip out the cubes into piles. Ask volunteers to count them and make them into sticks.

Make a table on the board using the data from the sticks.

Story	Votes
The Friendly Giant	6
Rainforest Stories	5
My Dog Peach	3

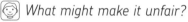

Ask the class questions about this data in order to establish what it tells them.

Give the 'talking stick' to children who want to answer your question or comment on what another child has said.

Reordering the information
You may want to redraft the table, with the stories in order of votes: favourite first, next favourite second, and so on.

There were six cubes in the 'Friendly Giant' pot. Where do I write the '6'?

How many people wanted the story about Peach, the dog?

Which story had most votes? How can we tell?

What do the numbers tell us? Anything else?

Do you agree with what Avtar just said? Why not?

So, was this fair? What do you think?

Tell your neighbour how we used voting today.

Support: Choose between two stories only.

Extend: Children decide which books to vote on and collect the data themselves.

Plenary

Read the chosen story.

Assessment for learning

Can the children	If not
Say what the numbers on the table mean?	Talk through with the children how you put cubes in pots, counted them and made the table, emphasising why each step was necessary for the overall result.
Use the opportunity of holding the 'talking stick' to say whether or not they agree with another child's idea, and why?	Provide more opportunities for children to express ideas and for others to comment on them. Use the 'talking stick' technique (p10) regularly so that children feel comfortable with it.
Say how voting helped choose a story to read?	Encourage children to talk about what would happen if lots of people all wanted different stories and nobody could agree. Acknowledge that if you only read the story that got the most votes, quite a few people will not get what they want.

Organising a table 2
Classroom technique: Talking partners

Learning objectives

(m) Maths
Make a table: sort and organise information

Speaking and listening
'Join in a discussion with a small group'
Contribute to small-group discussion

Personal skills
'Discuss and agree how to work'
Work with others: discuss and agree ways of working

(w) Words and phrases
count, sort, group, table, chart, collect, order, most, fewest, least

(r) Resources
for each group:
pots of coloured counters or natural objects (things that can be sorted into three or four categories)
copies of RS22 (prepare the tables according to the sorting material)

Introduction

Revise the process of making a table. Collect data about the colours of the children's tops (or type of shoes/favourite colours) and involve them in recording it on a table.

Colour of top	Number
red	
black	
grey	
blue	

Briefly discuss the data.

(m) *Am I putting this number in the right place? Where should it go?*

(m) *What do these numbers tell us?*

Groups of three

Children work in groups of three. Give each group a pot of about 60 objects that can be sorted, for example, by colour. Children work together to sort them, count how many there are of each kind and fill in the chart on RS22.

(m) *What do you need to write in these boxes?*

(m) *What do you need to do now you've counted your buttons?*

(s) *Do you agree with what Jason said? Why not?*

(s) *Tell Kylie where you think she should write her number. Explain why.*

(s) *How are you deciding who will do what? Is that fair?*

(s) *Is it a good idea if one person fills in the whole table? Why do you think that?*

Sharing the task
Remind children that they must make decisions as a group: who will count which set, how to organise filling in the table, and so on. Emphasise that you are looking for a fair sharing of the work, good manners and thinking about others.

Support: Limit the number of objects.

Extend: Don't prepare the table – let children do this themselves.

Plenary

Ask questions about the groups' results and tell children to discuss these
in their groups.

Which set was there most/fewest of? How many were there in that set?

Were there more than ten of any object? More than 20?

Put the sets of buttons in order from most to fewest.

*If you counted all your things together, do you think you would have more
than 50? Why do you think that?*

Assessment for learning

Can the children

Fill in the table correctly and answer questions
about it?

Tell their group confidently what they think the
answer is to one of your questions?

Allow everyone in the group to take an equal part
in the task?

If not

Ask children to check their work and encourage
them to self-correct. Do more work with the class
on completing tables about data that interests
them (sport, stories, pop stars, and so on).

Give the group a 'talking stick' (p10) to pass round
and explain that the person with the stick has a
turn to talk and be listened to.

Do more activities using 'Rotating roles' (p10),
where the sharing of the task is formalised by
assigning each person a specific role.

Interpreting data
Classroom technique: Devil's advocate

Learning objectives

Maths
Interpret data

Speaking and listening
'Reach an understanding with your partner'
Reach a common understanding with a partner

Personal skills
'Develop confidence about what you think and decide'
Improve learning and performance: develop confidence in own judgements

Words and phrases
count, sort, group, set, collect, chart, order, most, fewest, least

Resources
four or five sheets of A3 paper
for each pair:
sticky tape
strips of paper
glove puppet (optional)

Introduction

Briefly discuss what sports or outdoor games and activities children enjoy. Write the names of four or five of these on the A3 sheets of paper.

swimming	skipping	riding bikes	football

Explain to the children that they are going to choose their favourite activity and put their name on that sheet of paper.

Tell your partner which activity you are going to choose.

Pairs

Give each child a strip of paper on which to write their name. Children come to the front, one at a time, and stick their name on one of the sheets.

Make some statements about the data – some true, some false.

Pairs discuss whether or not your statement is true, aiming to reach agreement. They also discuss how they know the correct answer.

They then tell the class what they said to their partner and establish the truth of each statement.

Ordering the sets
Encourage the children to help you order the sets according to the number of names in each set.

Devil's advocate
When making false statements, use the glove puppet or put your hands behind your back and tell the children that you may have your fingers crossed.

(m) *I wonder if there will be more people who like swimming or more who like riding bikes. What do you think?*

(m) *What do all these people have in common?*

Tell your partner which set was chosen by only one person.

Six people chose football. Am I right? How do you know that?

Nobody chose skipping. True or false? And how do you know that?

🙂 *This set was chosen by the fewest people. Is that right?*

swimming
Tunde
Katie
Jordan
Luke

Support: When the charts are complete, write the number of names on each one.

Extend: Children make their own true or false statements in front of the class, for the other children to discuss.

Plenary

Talk briefly about the notion of confidence. Children turn to their partner and say how confident they feel about agreeing or disagreeing with your (or the puppet's) statements.

🙂 *Was it easy for you to decide whether the puppet was correct or not?*

🙂 *When the puppet said something that wasn't true, did it feel OK to say so?*

🙂 *Sometimes we aren't sure about something, and sometimes we are sure we are right. Did you know today when you were right?*

Assessment for learning

Can the children

ⓜ Decide whether a statement interpreting the data is true or not?

🗨 Tell you whether or not they agree with their partner?

🙂 Stick by an opinion even when you say, "Are you sure?"

If not

ⓜ Give the children buttons, counters or small-world objects to sort and talk to them about the results. This will help them gain more experience of sorting objects into clearly defined sets and counting the sets they have made.

🗨 Invite both children in the pair to tell you what they think and then ask each of them if they agree with the other.

🙂 Make the puppet say a simple true statement and ask if the children agree with it. Then repeat this with an obviously false statement and congratulate them on disagreeing with the puppet.

Self and peer assessment

Lesson 13: Sorting numerals	I think	My partner thinks
(m) I can sort numerals into two sets.	😊 🙁	😊 🙁
🗣 I listen to what my partner says.	😊 🙁	😊 🙁

Lesson 14: Organising a table 1	I think	My partner thinks
(m) I know what the numbers in the table mean.	😊 🙁	😊 🙁
🗣 I can tell the class what I think.	😊 🙁	😊 🙁

Name _____

Self and peer assessment

Lesson 15: Organising a table 2	I think	My partner thinks
(m) I can help sort objects and fill in the table.	😊 ☹️	😊 ☹️
I tell my group what I think the answer to a question is.	😊 ☹️	😊 ☹️

Lesson 16: Interpreting data	I think	My partner thinks
(m) I know what is true or false when I look at the charts.	😊 ☹️	😊 ☹️
I talk with my partner about the questions.	😊 ☹️	😊 ☹️

Measures

Learning objectives

	Lessons			
	17	**18**	**19**	**20**
ⓜ Maths objectives				
use simple measuring equipment	●			
measure using uniform non-standard units		●		
compare lengths by direct comparison			●	
read the time on an analogue clock				●
Ⓢ Speaking and listening skills				
contribute to small-group discussion	●			
talk about shared work with a partner		●		●
speak confidently in front of the class			●	
☺ Personal skills				
work with others: show awareness and understanding of others' needs	●			
work with others: work cooperatively with others		●		
improve learning and performance: critically evaluate own work			●	
improve learning and performance: reflect on learning				●

About these lessons

Lesson 17: Comparing weights

(m) Use simple measuring equipment

Young children sometimes have problems understanding that it is the heavier of two objects that pushes the balance pan down. Linking this with their own experience of holding and carrying objects can help them make that connection.

Contribute to small-group discussion

Classroom technique: Rotating roles

Working in a group and taking turns allows each child a space to express their opinion. You can encourage children to challenge each other's opinions and say why they disagree with them.

Work with others: show awareness and understanding of others' needs

Children enjoy working in a group, and it is an effective way to get them thinking about each other and what they need. Talking about what other people need raises children's awareness and gives value to empathic behaviour.

Lesson 18: Estimating capacity

(m) Measure using uniform non-standard units

Children find out how many mugs can be filled from a jug, in the process dealing with ideas such as whether it matters if some mugs are not completely filled and whether the amount poured out will fit back in the jug again.

Talk about shared work with a partner

Classroom technique: Tell your partner

This technique is used in both the introduction and the plenary. When you pose a question to the class, ask each child to turn to a neighbour and tell them the answer. This way, everybody is involved in answering questions.

Work with others: work cooperatively with others

The group task can only be fulfilled if children cooperate with each other.

Lesson 19: Estimating and comparing lengths

(m) Compare lengths by direct comparison

The idea that two lengths (or other measurements) can only be compared directly if they share a common base line is important and underpins much of measuring and data handling. This activity focuses on this crucial aspect of measurement.

Speak confidently in front of the class

Classroom technique: Heads or tails

Children discuss their task, knowing throughout the lesson that either one of them may be asked to talk about their work in the plenary. This encourages them to share responsibility for, and understanding of, the task.

Improve learning and performance: critically evaluate own work

Children talk with their partner (and perhaps the class) about their work and discuss whether they think they are getting better at estimating lengths. This discussion can help develop the habit of self-reflection.

Lesson 20: Reading time on a clock

(m) Read the time on an analogue clock

The way in which the two hands of a clock move at different speeds can confuse children. Supported by the teacher, this activity allows them to engage with clocks and their way of working.

Talk about shared work with a partner

Classroom technique: Peer tutoring

Children often understand explanations given by another child more readily than the same explanation given by an adult. In this activity, confident children become 'Experts' and help their peers understand an idea or a concept. This technique can benefit both children.

Improve learning and performance: reflect on learning

In the plenary, children talk to their partners about the lesson and say what they think they have learned.

Comparing weights
Classroom technique: Rotating roles

Learning objectives

(m) Maths
Use simple measuring equipment

Speaking and listening
'Join in a discussion with a small group'
Contribute to small-group discussion

Personal skills
'Think about what other people need'
Work with others: show awareness and understanding of others' needs

(w) Words and phrases
weigh, weight, balance, scales, heavy/light, heavier/lighter, heaviest/lightest, compare, guess, estimate, predict, record, explain

(r) Resources
two similar bags, one full of objects and one empty
for each group:
conkers (or equivalent)
pots labelled 'lighter than the conker' and 'heavier than the conker'
balances
objects such as table tennis balls, small toys, cubes, feathers, leaves

Size and weight
Young children may not see a distinction between size (volume) and weight, so make sure to include both large lightweight objects and small heavy ones. Then discuss what happens when they are placed on the balance.

Down and up
You may need to help children understand that when working with balances, 'down' means 'heavier' and 'up' means 'lighter' (some children may equate 'up' with 'big').

Judging weights
Encourage children to think for themselves and not be influenced by what the others say – and also to challenge each other if they disagree.

Introduction

Let children hold the two bags – one full and one empty. Talk about how the full bag pulls down the arm holding it.

Show the children two pots labelled 'lighter than the conker' and 'heavier than the conker', as well as a conker and another object such as a table tennis ball. Demonstrate how to 'weigh' the objects in your hands and let some of the children do the same.

Invite predictions from the children about what will happen when you put both objects in the balance pans. Then do this and establish what you have found out. Place the object in the appropriately labelled pot.

Repeat this process with other objects and other children.

(m) *When you are holding the full bag, does it pull your arm down? More than the empty bag? Why is that?*

(m) *Which object will push the pan down most? Why do you think that?*

(m) *The dinosaur pan is going down. Does that mean the dinosaur is heavier or lighter than the conker?*

(m) *This marble stayed up in the air. So which pot does it go in?*

Groups of three

Children work in groups of three. Give each group a set of scales, a conker, a selection of objects and two pots labelled 'lighter than the conker' and 'heavier than the conker'.

Child A chooses an object, and the group take turns to hold it and the conker and make a prediction out loud to the group.

Child A then places the object and the conker on the balance, and the group agree what they have found out and which pot the object belongs in. Child A puts the object in that pot.

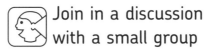
Child B now has a turn, and so on.

Do you all agree with Craig that the pencil is heavier than the conker? Tell him what you think.

Does that key belong in that pot? Tell your group why not.

When it's your turn to feel the objects, do you like having enough time to do it properly?

Support: Work with this group.

Extend: Ask children to find a way to record their work.

Plenary

Work with the class to show and compare the contents of their pots and discuss what can be deduced from this.

What can be said about all the objects in one pot?

Why might one group have a marble in the 'heavier' pot and another group have one in the 'lighter' pot?

Do any 'lighter' pots contain objects larger than the conker?

Are the things in the 'heavier' pot all bigger than the conker? Why is that?

Thinking about others
Have a mini-plenary partway through the activity and talk about what people working together in a group need.

Assessment for learning

Can the children

Find out whether an object is heavier or lighter than the conker?

Say why they do or don't agree with another child's estimate?

Give other group members time to feel the objects and make their decisions?

If not

Make balances and a range of objects available for free play. Play at 'human balances', using carrier bags held in each hand and filled with shopping items.

Check with another activity if the maths level is too high for the child. If children are daunted by the idea of working in a group, do more paired activities involving 'Tell your partner' (p10).

Use a 'My turn' label which the children pass round the group. The child with the label must be respected and given as much time as they need before passing it on to the next child.

Estimating capacity

Classroom technique: Tell your partner

Tell your partner
At each stage, the children turn
to their neighbour and take turns
to tell them what they think,
and why.

Learning objectives

Maths
Measure using uniform non-standard units

Speaking and listening
'Talk about your work with your partner'
Talk about shared work with a partner

Personal skills
'Work well with others'
Work with others: work cooperatively with others

Words and phrases
full, half full, empty, holds, container, guess, predict, estimate, describe, explain, record

Resources
for each group:
copy of RS23
tray
clear plastic jug full of rice/beans/small toys/water/juice
clear plastic mugs, cups or pots (identical)
funnel (optional)

U&A Making predictions
Encourage the habit of predicting to help children develop as mathematical thinkers.

Changing estimates
This kind of 'cheating' has educational value, whereas filling in 'estimates' after the pouring is all done has none.

Working cooperatively
Stop the groups for a mini-plenary and discuss ways of working. What behaviour makes the group task easier (for example, listening to each other, taking turns, sharing)? What behaviour makes it more difficult (for example, grabbing, ignoring each other, messing around)?

Introduction

Show the children a clear plastic jug full of rice, beans or similar and some clear plastic mugs.

Ask the class how many mugs they think can be filled from the jug. Collect in a few children's estimates and record them on the board.

Start filling the mugs. When you are about halfway, stop and discuss the estimates. Give the class an opportunity to have a second guess.

How many mugs?		
Name	**1st guess**	**2nd guess**
Liam	3 mugs	5 mugs
Jon	7 mugs	6 mugs

Then complete the filling of the mugs.

Finally, pour the contents of the mugs back into the jug. This time, ask for predictions as to whether the jug will be full or not full or whether it will overflow.

Tell your partner if the mug is full enough. Why does it matter?

Tell your partner whether you want to change your guess.

When you talk to your partner, what do you want them to do to show they are listening? What would you not like them to do?

Groups of three

Children work in groups of three. Give each group a tray with a clear plastic jug full of rice, beans or similar, some clear plastic mugs and a copy of RS23.

The groups then discuss how many mugs they think will be filled and to record their estimates on RS23.

The children then fill one mug and look at how much the level in the jug has gone down. Each child decides at this point whether to make a second guess.

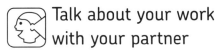
The children complete the filling of the mugs and record the actual result.

(m) *How full will you fill that mug? Does it matter?*

(m) *What about the beans you spilled in the tray? Where do they belong?*

(☺) *How will you share the work? Tell me what you will each do.*

(☺) *Can you tell me what your group needs to do?*

(☺) *Tell Katie what you need to do with this record sheet.*

Support: Work with this group.

Extend: Children fill the mugs first and predict whether their contents will fit into the jug. Then repeat the activity with the same jug but mugs of a different size.

Plenary

Repeat the activity from the introduction, using the same jug but larger (or smaller) mugs. Discuss whether the jug will fill the same number of mugs, or more, or fewer.

(☺) *These mugs are smaller than the ones I used before. But the jug is the same. Will I fill more mugs, or fewer mugs, or the same number? Tell your partner what you think.*

Is it fair?
Children need to think about being 'fair'. Are their results valid if the glasses are all filled to different levels?

Tell your partner
Make use of this technique again to keep children engaged and talking.

Assessment for learning

Can the children

(m) Fill the mugs carefully and completely, understanding that spilling some of the jug's contents will spoil the measurement?

(☺) Tell you what their partner has just said?

(☺) Share out the work of pouring from the jug?

If not

(m) Provide opportunities for real-life work with capacity, as children need to experience sand and water play and to talk about what they are doing with adults and peers.

(☺) Go over the task a stage at a time and each time ask children to tell you what you have just said.

(☺) Talk about the value of sharing and give children more opportunities for similar work with capacity, as they may have had little of this kind of experience.

Estimating and comparing lengths

Classroom technique: Heads or tails

Learning objectives

m Maths
Compare lengths by direct comparison

Speaking and listening
'Speak in front of the class'
Speak confidently in front of the class

Personal skills
'Evaluate your own work'
Improve learning and performance: critically evaluate own work

w Words and phrases
length, long, short, longer, shorter, longest, shortest, far, near, close, about the same as, compare, measure, guess, estimate

r Resources
two long strips of paper (about 1 m) of slightly different lengths for each pair:
several strips of ribbon or paper/sticks/straws of various lengths (20 cm to 1 m)
spinner showing 'longer' and 'shorter' (or use RS24 and a paper clip)
counters

A common base line
To emphasise the need for a base line, ask one child to stand on a stool while holding their strip up and the other child to stand close by with theirs. Discuss whether, in this situation, you can know for sure which strip is longer.

Agreeing an estimate
Children are committed to an estimate by agreeing this with their partner. Emphasise that they must make this decision together.

Heads or tails
Explain that either partner may be called on to talk about their experience, so they need to help each other prepare for this. Support their discussion with questions such as: "When was estimating easy? When was it hard?"; "How do you check which strip is longer?" Then use these same questions in the plenary.

Introduction

Ask two children to stand some distance apart, each holding up a long strip of paper (about 1 m) by its end. Discuss with the class how to compare the lengths of the strips.

Establish the method of placing the strips together with a common base line.

m *Does it matter if they hold the strips like this* [with no common base line]*? Why does it matter?*

Tell us which strip is longer and which is shorter. How can you be sure?

Pairs

Give each pair of children several strips of paper and some counters. Each child chooses one of the strips of paper and holds it up an arm's length away from their partner's. The children discuss and agree an estimate about which strip is longer before bringing them together and checking. If their estimate was correct, both children take a counter.

They then spin the spinner to find out which of them will win another counter: if the spinner lands on 'shorter', the child with the shorter strip wins an extra counter; if it lands on 'longer', the child with the longer strip wins an extra counter.

Before the plenary, pairs count their counters to find the 'winner'. Give them a minute or two to talk about how easy or difficult they found the estimating and whether they got better at it as they went on.

m *How do you know that this strip is longer than that one?*

When did you find it easy to estimate which strip is longer?

Do you think your estimation skills are improving?

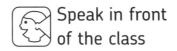
Support: Make sure the strips are of distinctly different lengths. Work with this group.

Extend: Give children strips of various widths and lengths and talk about whether the width affects the estimation and comparison of length

Plenary

Speaking to the class
Note which children talk to the class. On another occasion, ask others to talk about another piece of work so that, over time, every child has experience of this.

Ask a pair to come to the front of the class and toss a coin to decide which child will speak. Help the child talk about how they found the estimating task.

Repeat this with other pairs.

How do you check which strip is longer?

Make sure you speak loudly and clearly so that everyone can hear you.

How hard did you work at this task?

Pat yourself on the back if you think you have worked well today.

Assessment for learning

Can the children

(m) Check accurately which strip is longer by holding the strips so that they have a common base line?

Talk clearly to the class so that everyone can hear them?

Accurately assess how focused they have been on the task?

If not

(m) Focus on length over a period of time, in its various everyday uses: making belts for teddies, comparing lengths of skipping ropes, discussing whether a straw will reach to the bottom of a bottle, and so on. Then try this activity again.

Ask children to speak just to you and relay their words to the class yourself. This way, the child will still have the experience of communicating with everyone, but without the need to speak out.

Give the child feedback yourself. Encourage other children to give feedback.

Reading time on a clock

Classroom technique: Peer tutoring

Learning objectives

m Maths
Read the time on an analogue clock

Speaking and listening
'Talk about your work with your partner'
Talk about shared work with a partner

Personal skills
'Think about what you have learned'
Improve learning and performance: reflect on learning

W Words and phrases
time, day, morning, afternoon, evening, night, takes longer, takes less time, hour, o'clock, half past

r Resources
geared clock or display equivalent for the teacher
display copy of RS25
spinner made from RS26 and a paper clip
for each pair:
geared clocks
(if available) or other clocks with moveable hands (if necessary, these can be made from RS27 with a paper fastener)

Which clocks?

Real clocks from jumble sales and the geared clocks you have in school are a great help for young children: when the minute hand is moved, the hour hand moves itself to match. If real clocks are not available, children will need to move both the minute hand and hour hand. You could ask each partner in a pair to be responsible for moving one of the hands.

Peer tutoring

Identify any pairs that seem to be having problems. Readjust the pairs so that less confident children work with a confident partner. Name the confident children 'Clock Experts' and give them the task of supporting and teaching their partners – *not* doing the work for them.

Introduction

Display RS25 to the class, showing Giant's bed- and mealtimes. Also display a clock set to 7 o'clock: Giant's getting-up time. Give each pair of children a clock with moveable hands, also set to 7 o'clock.

Tell a story about a giant who is so busy with her garden, animals and friends that she sometimes misses her meals.

Spin the spinner to decide how long Giant spends with her garden, animals and friends when she gets up. Pairs move their clock hands on accordingly, so they show the new time, and hold up their clocks for you to see.

Move on your own clock, too, as a check. Now tell the class to look at the chart on RS25 to see whether Giant is in time for her meal or whether she has missed it.

Continue spinning the spinner and moving on the clocks to see how many meals Giant manages to eat and how many she misses.

m *The minute hand and hour hand move differently. Which one moves faster?*

m *How can you tell which one is the minute hand?*

Talk *to your partner about whether it always means the same thing when the minute hand points to 12.*

Clock *Experts, explain to your partners how you can tell what time the clock says.*

Are *you ready to be a Clock Expert? Are you sure enough about this work?*

Pairs

Repeat the activity. This time, pairs of children predict how many meals and snacks they think Giant will get. Record a range of predictions for the children, but do not record any names to indicate who said what.

At the end, compare the meals Giant managed to eat with the predictions.

What is the maximum number of meals and snacks that Giant managed to eat this time?

 Might she miss every meal? Why do you think that?

 Clock Experts and partners, talk together about how to move the hands on half an hour.

 Tell me what you think you are learning about in this lesson.

Support: If only a few real geared clocks are available, make sure these are given to the children who require more support. Explain that they only need to move the minute hand.

Extend: On another occasion, children repeat the activity in pairs, with a spinner and geared clock, but no adult intervention.

Peer tutoring

Again, identify children who still seem to be having problems. On another occasion, set up groups of three to repeat the activity, with one confident child acting as 'tutor' and using the spinner in the way that you did. The tutor does not touch the other children's clock, but explains to them what to do.

Plenary

Tell everyone to set their clocks to 12:00. Each pair moves the hands on one hour (or half an hour) at a time and holds up their clocks. Each time, ask one or two children what time is now shown. Repeat this process, then ask children to talk to their partners about what they have learned today.

 Tell your partner what today's lesson has been about.

 You learned about telling the time. What did you learn about listening?

 Did you find anything difficult today? Tell your partner what it was.

Assessment for learning

Can the children

 Read any hour or half-hour on an analogue clock?

 Talk with, and listen to, their partner about the clock activity?

 Identify one thing they have been learning about in the lesson?

If not

 Set an alarm for various hours (for example, 10:00 or 2:30) and stop the class to look at the clock when the alarm goes. Ask children to draw how the clock looks when the alarm goes.

 Do more activities involving 'Tell your partner' (p10) in which children formally take turns to speak to their partner.

 Make a point of telling children what they will be learning about in each lesson. Regularly ask partners to tell each other what they have been learning about.

Name _____

Self and peer assessment

Lesson 17: Comparing weights	I think	My partner thinks
(m) I can find out whether something is heavier or lighter than the conker.	😊 ☹️	😊 ☹️
I can tell my group what I think.	😊 ☹️	😊 ☹️

Lesson 18: Estimating capacity	I think	My partner thinks
(m) I can find out how many mugs can be filled from a jug.	😊 ☹️	😊 ☹️
I listen carefully when my partner talks to me.	😊 ☹️	😊 ☹️

Name _____

Lesson 19: Estimating and comparing lengths	I think	My partner thinks
(m) I can find out which of two strips of paper is longer.	☺ ☹	☺ ☹
👤 I talk clearly to the class so that everyone can hear me.	☺ ☹	☺ ☹

Lesson 20: Reading time on a clock	I think	My partner thinks
(m) I can tell the time when it is 'o'clock' or 'half past'.	☺ ☹	☺ ☹
👤 I talk to my partner and also listen to what they say.	☺ ☹	☺ ☹

Self and peer assessment

Shape and space

Learning objectives

	Lessons			
	21	**22**	**23**	**24**
Ⓜ Maths objectives				
use everyday language to describe positions	●			
sort familiar 2D shapes		●		
describe 2D shapes and 3D solids			●	
make and describe patterns				●
Ⓢ Speaking and listening skills				
reach a common understanding with a partner	●			
talk about shared work with a partner		●		
listen to others and ask relevant questions			●	
give accurate instructions				●
☺ Personal skills				
work with others: work cooperatively with others	●		●	
improve learning and performance: develop confidence in own judgements		●		
work with others: overcome difficulties and recover from mistakes				●

About these lessons

Lesson 21: Describing positions

(m) Use everyday language to describe positions

There is a great deal of positional vocabulary that is useful for young children to acquire. In this activity, children describe to a partner whereabouts on a picture to place a counter – which means they must think about how to describe one position among many possible ones.

Reach a common understanding with a partner

Classroom technique: Barrier game

Children give instructions about where they have placed counters on their picture to enable their partner to do the same on an identical picture. The listener may ask questions to check they are both in agreement about the position that is being described.

Work with others: work cooperatively with others

Children will only succeed at the shared task if they cooperate with each other.

Lesson 22: Sorting shapes

(m) Sort familiar 2D shapes

In this activity, children sort 2D shapes physically onto PE mats. To do this, they need to look at the shapes carefully and think about their properties. The game can be made easier or more challenging to suit the age and ability of the children.

Talk about shared work with a partner

Classroom technique: Tell your partner

To ensure that all children think and speak about the properties of their shapes, they are asked to turn to their partner and tell them about the shape.

Improve learning and performance: develop confidence in own judgements

Children look at the shapes in a set and decide what they all have in common. Inviting children to make and express judgements such as this, and accepting all ideas, helps them develop confidence – even if they are not always correct.

Lesson 23: Describing and naming shapes

(m) Describe 2D shapes and 3D solids

In this activity, children guess which shape or solid has a stamp hidden under it. They are not allowed to point at the shapes and solids, so have to describe which one they mean by using the language of shape.

Listen to others and ask relevant questions

Classroom technique: Talking stick

Children all take turns as a 'Hider', so everyone has a chance to be the one 'in the know'. The others are 'Guessers' who must ask questions to establish which shape is being described. The child whose turn it is to ask a question holds the 'talking stick', which signals to the rest of the group that they are to be quiet and listen to that child.

Work with others: work cooperatively with others

The success of the game depends on children cooperating with each other.

Lesson 24: Making patterns

(m) Make and describe patterns

Pattern underlies all mathematics. Once children can see a pattern – in numbers, shapes, a procedure – they have a secure basis on which to continue working and the mathematics can make sense to them. This lesson gives children practice in making patterns involving assorted counting objects.

Give accurate instructions

Classroom technique: One between two

Children share a set of counting objects and a task. The 'Doer' does only what they are told to by their partner, the 'Talker'. This encourages the Talker to think carefully about the instructions they give.

Work with others: overcome difficulties and recover from mistakes

Pattern making is a good activity in which to practise looking for and correcting mistakes, because inconsistencies in patterns tend to be obvious. Encourage children to check their work for themselves and, if they spot an error, to work out how to correct it.

Describing positions

Classroom technique: Barrier game

Learning objectives

(m) Maths
Use everyday language to describe positions

Speaking and listening
'Reach an understanding with your partner'
Reach a common understanding with a partner

Personal skills
'Work well with others'
Work with others: work cooperatively with others

(w) Words and phrases
position, over, under, underneath, above, below, on, in, in front, beside, before, after, next to, opposite, between

(r) Resources
display copy of RS28
for each pair:
two copies of RS28
coloured counters
copies of RS29 and
cubes (optional)

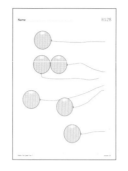

Modelling language
This is an opportunity to remind children of the kind of positional language they can use when they come to do the activity in pairs: "Two of the balloons are touching. Put a red counter on the balloon just below."

Back to back
Some children may be able to work sitting back to back. Others may prefer to use a large book or sheet of card as a barrier.

U&A Using mental imagery
Activities such as this encourage and support children to use their mental imagery skills in a mathematical context.

Introduction

Children work in pairs. Briefly display RS28 to the children, then give each pair two copies of RS28 and some counters.

Choose a coloured counter to place on a balloon on your sheet. Describe the colour and position of the counter to the children and tell them to put a counter of the same colour on the same balloon on their copies. Next, compare and discuss whether all pairs have the counter in the same position.

Continue until you have placed six counters, one on each balloon.

(m) *Where did I say to put the red counter? If yours is not in the correct position, move it there now.*

(m) *Mel, look at my sheet. Where shall I tell the class to put their next counter?*

(Speaking) *Partners, do you both agree where the counter belongs? If not, I'll say it again and you can talk about where you think I mean.*

Pairs

The pairs now work with their copies of RS28 and counters. Children sit so that they cannot see their partner's work.

Pairs repeat the activity from the introduction. When their sheets are complete, they compare their work. If there are any discrepancies, they correct these.

They then swap roles and repeat the activity.

(m) *When Ryan tells you where he has put his counter, what do you see in your head?*

(Speaking) *How do you check whether your counters are in the same place?*

(Speaking) *If you are not sure which balloon Gary wants you to put your counter on, what can you do?*

(Personal) *When it's your turn to do the telling, how can you help Aysha?*

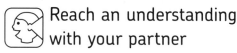
Support: Children check the position of each counter after it is placed, rather than waiting until all six are in position.

Extend: Give children copies of RS29 and nine coloured cubes or counters.

Extend

You could even give children a 3 × 3 grid with no pictures. Children will need to give quite sophisticated instructions: "Put a red cube on the top row, on the left."

Plenary

Ask the class to sit on the carpet. Invite one child at a time to go and stand in the position you describe: "Amy, stand beside the door, in front of the bookcase."

Tell your partner whether you think Robert is standing where I asked him to.

Assessment for learning

Can the children	**If not**
(m) Use appropriate vocabulary to describe position?	(m) Note any words that are not being used and plan teaching to focus on the vocabulary and ideas.
Ask for further information when they don't understand their partner's description?	Set up activities where the onus is on children to ask for information from you or from each other, rather than to wait passively until they receive it.
Deal patiently with their partner if they need help?	Remind children that success in this task is a joint responsibility and that they must work together to share their ideas.

Sorting shapes

Classroom technique: Tell your partner

Learning objectives

(m) Maths
Sort familiar 2D shapes

Speaking and listening
'Talk about your work
with your partner'
Talk about shared work with
a partner

Personal skills
'Develop confidence
about what you think
and decide'
Improve learning and
performance: develop
confidence in own judgements

(w) Words and phrases
circle, triangle,
square, rectangle, side,
edge, corner, sort, organise,
same, different

(r) Resources
flat plastic shapes
two PE mats
chalk
question cards cut
from RS30

Tell your partner
Children take turns to say
one thing about their shape.
Orchestrate this by assigning
each child in a pair the role of
either a 'Lion' or a 'Tiger'. Ask the
Tigers to say something. Then,
after ten seconds or so, ask the
Lions to say something different.

Tell your partner
Occcasionally, children turn
to their partner and say what
they think might be on the
question card.

Introduction

Do this activity in a large space such as the school hall.
Label the PE mats 'Yes' and 'No', using a piece of chalk.
The children form pairs and sit down near – but not on
– the mats.

Give each pair a 2D shape. Pairs tell their partner one
or two things about it.

Hold up one of the question cards and explain that you
are the 'Gatekeeper' and that you are going to tell the
children which mat to go to. Ask a child to bring their
shape to you and then ask them the question on the
card. According to their answer, they go and sit on the
appropriate mat with their shape.

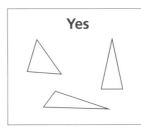
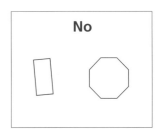

(m) *What is the same about all the shapes in this set?*

(m) *What is different about this set of shapes and
that one?*

(🗫) *Who has a shape with six sides? Tell your partner
how many corners it has.*

(🗫) *Is your shape still a triangle when you turn it
round a bit?*

(🗫) *Rectangles aren't all the same. If you have a
rectangle, tell your partner whether it is fat or
thin and whether it is big or small.*

Whole class

Start the activity again, giving each child a new shape.
This time, two children to join you. Together, the three
of you choose another question card in secret.

As each child holds up their shape, the three of you
decide what the answer is and direct the child to the
correct mat.

When everybody is seated on a mat, take suggestions as to the question on the card and establish what it actually says.

 *I think that too is a triangle. But tell me how you know.*

Tell us why you think our card says 'Is it a square?'

Tell me why this shape doesn't belong here.

Support: Less confident children suggest what is on the label before more assertive children have their say.

Extend: Children play a version of this game independently, using sheets of paper instead of mats.

Plenary

Ask each child, in turn, to come and remove a shape from the mats, using instructions such as "Take a shape with four corners"; "Take a curved shape"; "Take a shape that has not got straight sides."

Finally, children 'tell their partner' something about the properties of their shape.

 Come and take a triangle.

Take a shape that is not square.

Is that really a rectangle? How do you know?

Assessment for learning

Can the children

m Name familiar shapes and sort them from other shapes?

Turn readily to their partner and say something about their shape?

Say why they think a shape belongs in a particular set?

If not

m Have a 'Shape of the week' for a few weeks. Focus on its name and properties. Then try the sorting activity again.

Practise 'Tell your partner' (p10) activities in which children talk about more familiar subjects such as favourite toys, games or sports.

Confirm children's judgements, explaining how you know they are correct: "Yes, it does have four sides: 1, 2, 3, 4. Well done!"

Describing and naming shapes

Classroom technique: Talking stick

Describing shapes

Some children may find it quite challenging to name a shape and its colour or size if, for example, there are several squares. Others will be able to deal with more complex descriptions such as 'the shape with five sides'. Use similes to help the children – for example, 'the shape a bit like an arrow'. Choose sets of shapes and solids to suit the children.

A range of shapes

This is an opportunity for children to learn that some shapes only have one form (for example, cubes, squares and circles), while others have all sorts of forms (for example, triangles, cuboids, pyramids). Try to include a variety of shapes, rather than just one of each kind.

Introduction

Lay a selection of 2D shapes and 3D solids out on the desk in front of you and give each child a copy of RS30. The children close their eyes while you hide a stamp under one of the shapes.

Then children open their eyes and take turns to guess under which shape or solid the stamp may be. Explain that they must use description or naming, rather than pointing. Children use RS30 as a guideline.

Each time a shape or solid is eliminated, remove it from the set. Continue until the stamp is found.

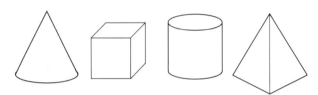

"Is it under the shape like a witch's hat?"

(m) *You asked, "Is it under the shape with five sides?" No, it isn't.*

(m) *It isn't under any of the shapes with a square face.*

(speaking icon) *What would be a useful question to ask next?*

Groups of three or four

Children do the same activity in groups of three or four, taking turns to be the 'Hider'.

The other children pass round a 'talking stick', which entitles them to ask one question about the whereabouts of the stamp.

(speaking icon) *Tell me which shape Kelly asked about. Yes, the half-moon shape. Do you know which one that is?*

(personal icon) *Can you suggest a question for Imran to ask?*

(personal icon) *Tell me one way that you can be helpful to the other people in your group.*

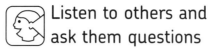
Support: Work with this group.

Extend: Encourage children to eliminate more than one shape or solid at a time, using categories: "Is it under any of the squares? If it isn't, remove all squares."

Plenary

Hide a stamp under a shape or solid as before. This time, give the children clues that eliminate several shapes or solids at a time: "It's not under a cube. So which solids can we take away?"

Jane asked, "Is it under one of the squares?" That's a useful question to ask.

Assessment for learning

Can the children	If not
Accurately name a cube, cuboid, pyramid, sphere, cone, cylinder, circle, triangle, square, rectangle?	Choose a new shape or solid each week to name, talk about, draw round and build with. Play more games involving shapes and solids (see also Lesson 22, p94).
Say which shape someone has just asked about?	Make sure that you and the children are articulating clearly and pronouncing shape words correctly. Encourage careful listening by playing games such as 'Simon says' (in which the leader says something and the children repeat it only if the phrase was preceded by the words 'Simon says...'). Also be alert for any hearing problems.
Wait patiently for their turn?	Suggest that children spend their waiting time planning which shape or solid they are going to ask about when it is their turn.

Name _____

Self and peer assessment

Lesson 21: Describing positions	I think	My partner thinks
(m) I can say where something is.	🙂 ☹️	🙂 ☹️
👤 I can give clear instructions to my partner.	🙂 ☹️	🙂 ☹️

Lesson 22: Sorting shapes	I think	My partner thinks
(m) I can describe a set of shapes.	🙂 ☹️	🙂 ☹️
👤 I can explain my ideas to my partner.	🙂 ☹️	🙂 ☹️

Name _____

Lesson 23: Describing and naming shapes	I think	My partner thinks
(m) I can describe a shape.	😊 ☹️	😊 ☹️
I listen to what my partner tells me.	😊 ☹️	😊 ☹️

Lesson 24: Making patterns	I think	My partner thinks
(m) I can make a repeating pattern.	😊 ☹️	😊 ☹️
I can describe my pattern to my partner.	😊 ☹️	😊 ☹️

Self and peer assessment

Resource sheets

The resource sheets should be printed from the PDF files on the accompanying CD-ROM.

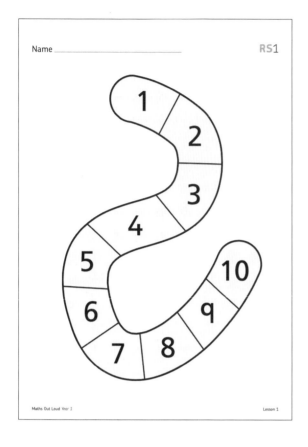

Name _____ RS1

Maths Out Loud Year 1   Lesson 1

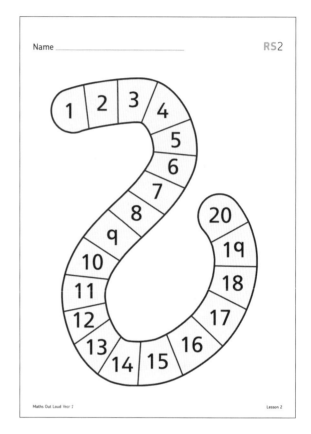

Name _____ RS2

Maths Out Loud Year 1   Lesson 2

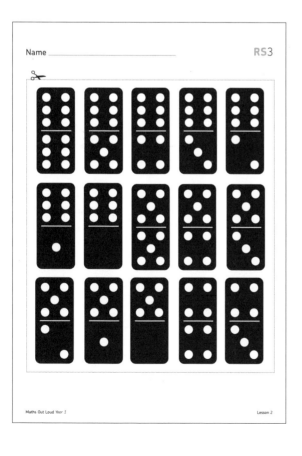

Name _____ RS3

Maths Out Loud Year 1   Lesson 2

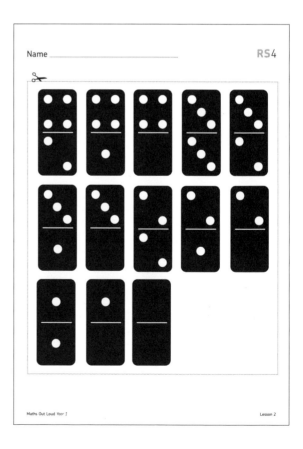

Name _____ RS4

Maths Out Loud Year 1   Lesson 2

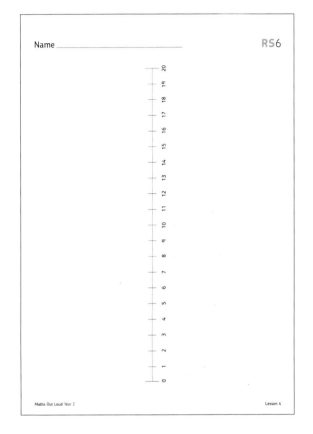

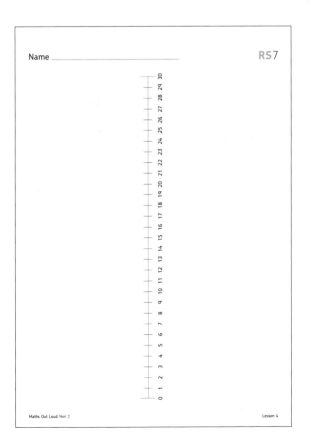

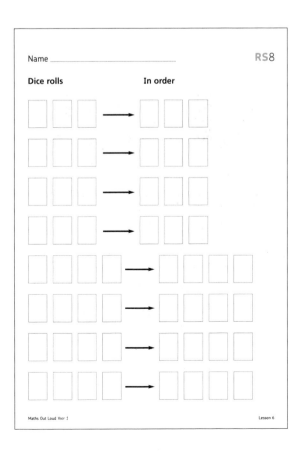

RS5 — Name _____

RS6 — Name _____

RS7 — Name _____

RS8 — Name _____
Dice rolls   In order

Maths Out Loud Year 1   Lesson 3
Maths Out Loud Year 1   Lesson 4
Maths Out Loud Year 1   Lesson 4
Maths Out Loud Year 1   Lesson 6

Name _____ RS9

Dice rolls   **In order**

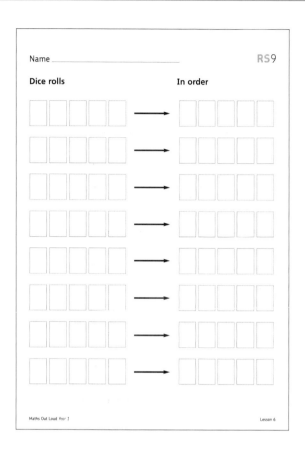

Maths Out Loud Year 1   Lesson 6

Name _____ RS10

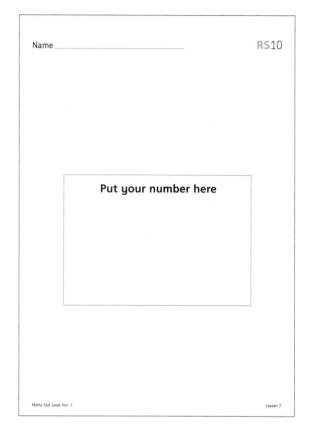

Put your number here

Maths Out Loud Year 1   Lesson 7

Name _____ RS11

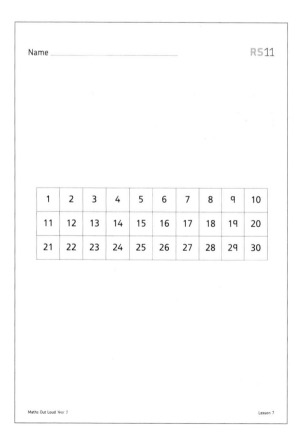

1	2	3	4	5	6	7	8	9	10
11	12	13	14	15	16	17	18	19	20
21	22	23	24	25	26	27	28	29	30

Maths Out Loud Year 1   Lesson 7

Name _____ RS12

1	2	3	4	5	6	7	8	9	10
11	12	13	14	15	16	17	18	19	20
21	22	23	24	25	26	27	28	29	30
31	32	33	34	35	36	37	38	39	40
41	42	43	44	45	46	47	48	49	50

Maths Out Loud Year 1   Lesson 7

Name _____ RS13

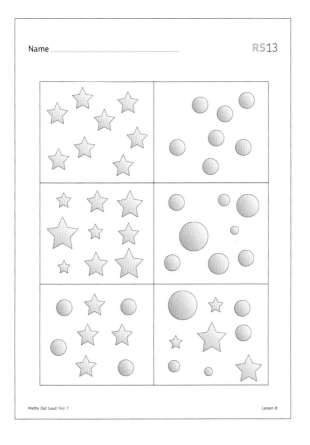

Maths Out Loud Year 1   Lesson 8

Name _____ RS14

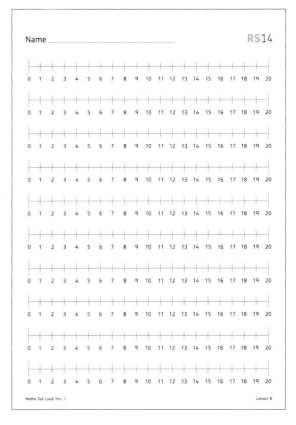

Maths Out Loud Year 1   Lesson 8

Name _____ RS15

Cubes I can see	Hidden cubes	

Maths Out Loud Year 1   Lesson 9

Name _____ RS16

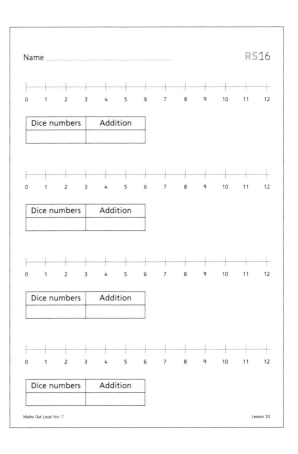

Maths Out Loud Year 1   Lesson 10

Name _____ RS17

| 0 | 1 | 2 | 3 | 4 | 5 | 6 | 7 | 8 | 9 | 10 | 11 | 12 |

| 0 | 1 | 2 | 3 | 4 | 5 | 6 | 7 | 8 | 9 | 10 | 11 | 12 |

| 0 | 1 | 2 | 3 | 4 | 5 | 6 | 7 | 8 | 9 | 10 | 11 | 12 |

| 0 | 1 | 2 | 3 | 4 | 5 | 6 | 7 | 8 | 9 | 10 | 11 | 12 |

Maths Out Loud *Year 1*   Lesson 10

Name _____ RS18

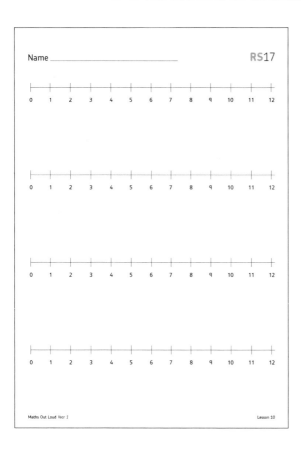

😊	👧	🐱	total	enough?
3	4	5	12	✓

Maths Out Loud *Year 1*   Lesson 11

Name _____ RS19

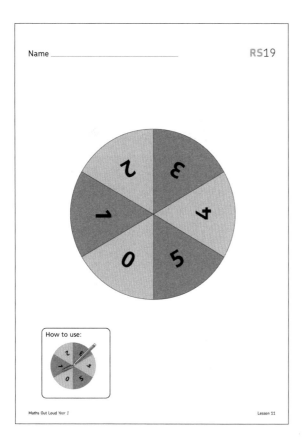

How to use:

Maths Out Loud *Year 1*   Lesson 11

Name _____ RS20

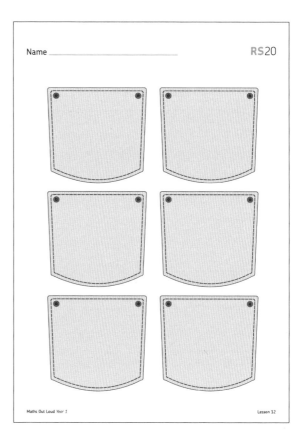

Maths Out Loud *Year 1*   Lesson 12

Name _____ RS21

0   1   2   3   4   5   6   7   8   9

Has it got any straight lines?

Has it got any curves?

Can you write this number without taking your pencil off the paper?

Has it got any corners?

Does it look the same upside down?

Does it look the same in a mirror?

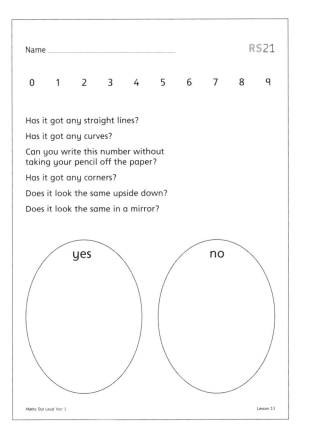

Maths Out Loud *Year 1*   Lesson 13

Name _____ RS22

Names . . . .

. . . .

. . . .

. . . .

	Number

Maths Out Loud *Year 1*   Lesson 15

Name _____ RS23

How many mugs?

Name	1st guess	2nd guess

Actual result . . . .

Maths Out Loud *Year 1*   Lesson 18

Name _____ RS24

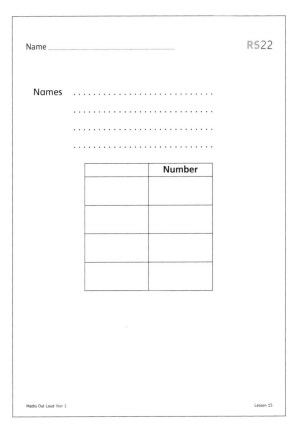

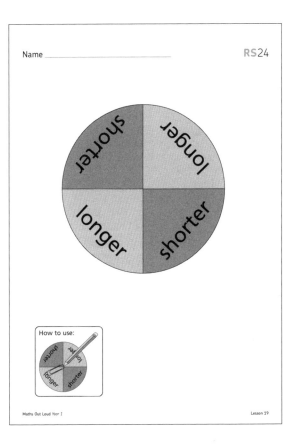

How to use:

Maths Out Loud *Year 1*   Lesson 19

Name _____ RS25

Giant's day

get up	7 o'clock
breakfast	8 o'clock
snack	11 o'clock
lunch	half past 12
tea	4 o'clock
dinner	half past 7
snack	9 o'clock
bed	10 o'clock

Maths Out Loud *Year 1* · Lesson 20

Name _____ RS26

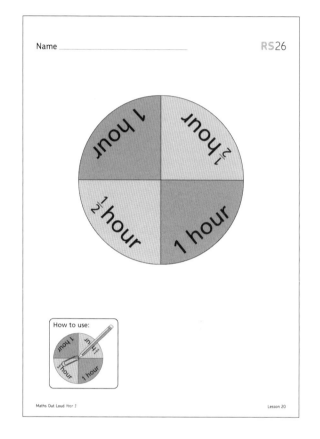

How to use:

Maths Out Loud *Year 1* · Lesson 20

Name _____ RS27

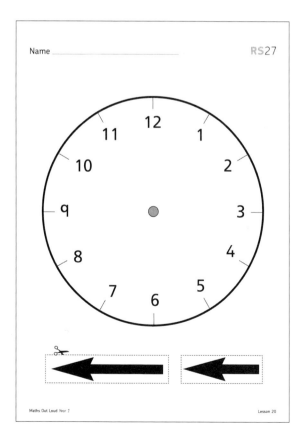

Maths Out Loud *Year 1* · Lesson 20

Name _____ RS28

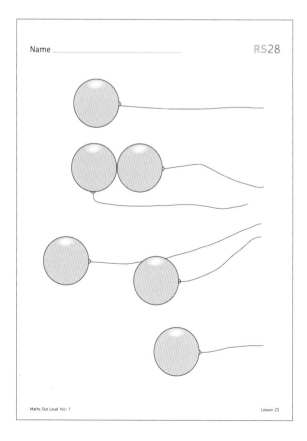

Maths Out Loud *Year 1* · Lesson 21

Name _____ RS29

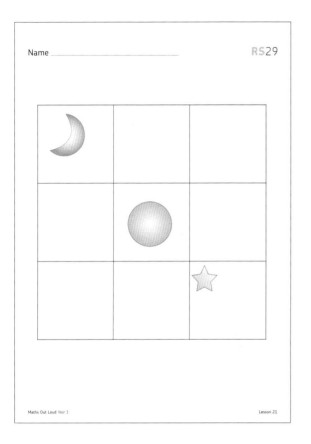

Name _____ RS30

| Is it a square? |
| Are all its sides the same length? |
| Is it a rectangle? |
| Are all its sides straight? |
| Is it a triangle? |
| Has it any curves? |
| Has it got 4 corners? |
| Has it got more than 4 corners? |
| Has it got 5 corners? |
| Has it got fewer than 4 sides? |
| Has it got 4 sides? |
| Is it a circle? |
| Has it got 6 sides? |